Objective Questions in 'A' Level Chemistry

Objective Questions in 'A' Level Chemistry

John Gunnell   Edgar Jenkins

Centre for Studies in Science Education,
University of Leeds

Oliver & Boyd

OLIVER AND BOYD
Tweeddale Court
14 High Street
Edinburgh EH1 1YL
A Division of Longman Group Limited

ISBN 0 05 002586 4

Filmset by Typesetting Services Ltd., Glasgow, Scotland

Printed by T. & A. Constable Ltd., Edinburgh

Contents

	Foreword	vii
	Preface	ix
	Directions for answering the questions	xi
1	Atomic Structure	1
2	Structure and Bonding	7
3	The Gaseous State	12
4	Energetics	18
5	Kinetics	24
6	Equilibria I	33
7	Equilibria II	40
8	The Periodic Table	47
9	The Chemistry of the Non-Metals	54
10	The Chemistry of the Metals	60
11	General Inorganic Chemistry	66
12	Hydrocarbons	71
13	Compounds of Carbon, Hydrogen and Oxygen	79
14	The Organic Chemistry of Nitrogen and the Halogens	86
15	General Organic Chemistry	93
16	Experimental Procedures	100
17	Revision Paper I	106
18	Revision Paper II	112

Foreword

There was a time, perhaps one or two decades ago, when objective testing was viewed with suspicion by many teachers of chemistry. Such suspicion nowadays is merely a confession of ignorance of the valuable role such tests can play in the teaching and assessment of chemistry. But it is only *one* method and care should always be taken to use a variety of methods and approaches. It would be tragic if the ready availability of such banks of objective questions misled teachers (or their pupils) into believing that the goal of advanced level teaching was merely to inculcate an ability to pass such tests. Indeed it could be argued that when a pupil has reached this stage he has merely mastered the vocabulary, grammar, and syntax of the language of chemistry. Whilst this in itself is a valuable intellectual exercise it is no more a study of chemistry than a similar mastery of Latin is a revelation of the subtle beauties and grandeur of the classics. It is a necessary prerequisite to the study of chemistry but we must in addition continually interweave references to the wider implications of these ideas. Chemistry is not only an exciting intellectual adventure, it is a study which has conferred great benefits on mankind, and its continued study and development holds the hope of solving many of the problems which beset both developed and developing countries today.

John Gunnell and Edgar Jenkins have done a great service in compiling this extensive set of pretested objective questions in advanced level chemistry. They have considerable experience of teaching at this level and of preparing chemistry graduates to become teachers, and this is reflected in the admirable balance and consistent standard of the questions.

I have great pleasure in commending this book to Advanced level pupils and to other students of chemistry. No one pretends the subject is easy but it can be immensely exciting and stimulating. The more practised you become in using the ideas and facts of chemistry the more fun you will have with the subject and the more valuable will be your personal contribution to the life of the community in which you live.

N. N. Greenwood
Professor of Inorganic and Structural Chemistry,
University of Leeds

Preface

Recent changes in the style and content of school chemistry teaching have been paralleled by corresponding developments in examination technique. In particular, the use of objective questions has enabled the examiner to sample a wide range of subject matter and a variety of thinking skills. For the science teacher, objective questions have become important resource material to be used in working with students.

This book provides a set of objective test papers in Advanced level chemistry. Four different item types are used: multiple choice, multiple completion, matching pairs (classification) and assertion-reason. Each paper has been pretested in schools by students following A-level chemistry courses and this has suggested that each paper can be satisfactorily attempted in approximately 45 minutes. Most of the questions have been designed to test more sophisticated skills than the ability to recall chemical information. As such, the papers will not be found easy and it is our hope that the questions will be found most useful as a contribution to the normal teaching–learning processes rather than simply as the basis of achievement tests.

We have tried to cross the traditional boundaries between the various branches of chemistry and have based the content on the principles to be found in most syllabuses in chemistry at Advanced level. We anticipate that the tests will be found useful by any teacher of chemistry at this level and by students who wish to deepen their understanding of the subject. For the convenience of teachers the answers have been published separately and in appropriate cases we have provided data or additional information which we felt might be of use in classroom discussion.

In the use of SI units we have been guided by the practice of the principal Examination Boards. Both I.U.P.A.C. and common nomenclature are given where this is necessary. It is assumed that students will have access to a simple periodic table and to a table of atomic masses.

We have drawn extensively upon the advice of colleagues and pupils in schools. We wish to acknowledge particularly the advice of Professor N. N. Greenwood, Dr. J. G. Broadhurst, Mr. P. J. Scott and the VIth form students and chemistry staff of the following:

Allerton Grange School; Bolling Grammar School; Cockburn High School; Cross Green School; Greenhead Grammar School; Hanson Grammar School; Harrogate High School; King's School, Pontefract; Lawnswood High School; Leeds Modern School; Queen Elizabeth Grammar School, Wakefield; Ramsden Technical College; Roundhay Girls' Grammar School; Tadcaster School; Temple Moor School; Wakefield Girls' High School.

John Gunnell
Edgar Jenkins
Centre for Studies in Science Education,
University of Leeds

Directions for answering the questions

See inside front flap for a summary of these directions.

Four different types of question are used in this book.

Multiple choice questions

In each of these questions a statement is followed by five alternative responses. Only *one* of these alternatives is correct. Indicate your choice of A, B, C, D or E on your answer sheet.

Multiple completion questions

In each of these questions, *one or more* of the responses given is correct. Consider each of the responses carefully and decide whether or not it is a correct answer to the question. With your conclusion in mind, indicate A, B, C, D or E on your answer sheet as follows:

A if only (*i*), (*ii*) and (*iii*) are correct
B if only (*i*) and (*iii*) are correct
C if only (*ii*) and (*iv*) are correct
D if only (*iv*) is correct
E if only (*i*) and (*iv*) are correct

Summarised directions for recording responses to multiple completion questions				
A	*B*	*C*	*D*	*E*
(*i*), (*ii*) and (*iii*)	(*i*) and (*iii*)	(*ii*) and (*iv*)	(*iv*)	(*i*) and (*iv*)

Matching pairs questions

In each group of questions, choose from the list A–E the responses which correctly answer each of the questions within that group. Note that each of *the letters A–E may be used once, more than once, or not at all.*

Assertion-reason questions

In each of these questions an assertion is followed by a reason. Consider the assertion and decide whether, on its own, it is a true statement. Consider the reason and decide if that is a true statement. If, and only if, you decide that *both* the assertion *and* the reason are true, consider whether the reason is a valid or true explanation of the assertion. Choose your answer as follows and indicate your choice on the answer sheet.

A If *both* the assertion and the reason are *true* statements and the reason *is a correct explanation of the assertion.*
B If *both* assertion and reason are *true* statements but the reason is *not a correct explanation of the assertion.*
C If the assertion is true but the reason is a false statement.
D If the assertion is false but the reason is a true statement.
E If both assertion and reason are false statements.

Summarised directions for recording responses to assertion-reason questions

	Assertion	Reason	Argument
A	True	True	Reason is a *correct* explanation of assertion
B	True	True	Reason is *not a correct* explanation of assertion
C	True	False	Not applicable
D	False	True	Not applicable
E	False	False	Not applicable

1 Atomic Structure

Multiple choice questions

1 What is the maximum number of atomic orbitals with the principal quantum number 3?

A 3
B 6
C 9
D 16
E 18

2 Which one of the following atoms in its ground state contains no unpaired electrons?

A lithium
B beryllium
C boron
D carbon
E oxygen

3 What fraction of the original mass of a radioactive substance of half-life 6 days remains undecayed after a period of 12 days?

A $\frac{1}{2}$
B $\frac{1}{4}$
C $\frac{1}{12}$
D $\frac{1}{36}$
E none of it

4 A radioactive isotope P has a half-life of 8 hours and an istope Q has a half-life of 12 hours. A mixture is prepared containing equal masses of P and Q. What is the mass ratio of P to Q remaining in the mixture after 24 hours?

A 1:1
B 1:2
C 2:1
D 2:3
E 3:2

5 Living organisms have mixtures of carbon-14 and carbon-12 that produce 15.3 ± 0.1 disintegrations of carbon-14 atoms per minute per gram of carbon. The half-life of carbon-14 is 5 600 years. Charcoal from Lascaux caves, France, the site of extensive cave paintings, produces 2.2 ± 0.1 disintegrations of carbon-14 atoms per minute per gram of carbon. What is the most probable age of the paintings?

A between 5 000 and 6 000 years
B between 11 000 and 12 000 years
C between 15 000 and 16 000 years
D between 18 000 and 19 000 years
E between 38 000 and 39 000 years

6 Which one of the following would *not* be used either to detect or to measure radioactive emission?

A a cloud chamber
B an electroscope
C a photographic film
D a scintillation counter
E a spectrophotometer

7 What is the number of protons in 2.24 litres of nitrogen gas at standard temperature and pressure?

A 0.60×10^{23}
B 4.2×10^{23}
C 6.0×10^{23}
D 8.4×10^{23}
E 8.4×10^{24}

8 Which one of the following elements has a first ionization energy which is greater than the first ionization energy of either of its horizontal neighbours in the periodic table?

A boron
B carbon
C nitrogen
D oxygen
E fluorine

9 For which one of the following elements is the ratio of the first ionization energy to the second ionization energy the lowest?

A sulphur
B chlorine
C krypton
D potassium
E calcium

10 For which one of the following compounds is the ratio of cation size to anion size greatest in value?

A CsF
B NaF
C LiF
D LiI
E CsI

11 Which one of the following sets of elements shows a decrease in atomic radius with increasing atomic number?

A the noble gases (He to Rn)
B the halogens (F to At)
C the alkali metals (Li to Fr)
D the alkaline earth metals (Be to Ra)
E the lanthanides (La to Lu)

12 A nuclide $^{222}_{86}Ra$ emits successively α, α, and β^- particles. What is the nuclide formed after these three emissions?

A $^{218}_{86}Ra$

B $^{214}_{83}Bi$

C $^{214}_{82}Pb$

D $^{214}_{81}Tl$

E $^{213}_{82}Pb$

13 The electron configurations below represent possible arrangements for a six-electron system.

(*i*) $1s^2 2s^2 p_x{}^1 p_y{}^1$
(*ii*) $1s^2 2s^1 p_x{}^1 p_y{}^1 p_z{}^1$
(*iii*) $1s^1 2s^2 p_x{}^1 p_y{}^1 p_z{}^1$
(*iv*) $1s^1 2s^1 p_x{}^1 p_y{}^1 p_z{}^1 3s^1$

Which one of the following is the order of *decreasing* potential energy of these systems?

A (*iv*) > (*iii*) > (*ii*) > (*i*)
B (*iii*) > (*iv*) > (*ii*) > (*i*)
C (*i*) > (*ii*) > (*iv*) > (*iii*)
D (*i*) > (*ii*) > (*iii*) > (*iv*)
E (*iii*) > (*ii*) > (*iv*) > (*i*)

Multiple completion questions

14 Which of the following species contain(s) 5 unpaired *d* electrons in its ground state?

(*i*) Cr
(*ii*) Mn
(*iii*) Fe^{3+}
(*iv*) Fe

15 Which of the following species is isoelectronic with Cu^+?

(*i*) Ni
(*ii*) Cu^{2+}
(*iii*) Zn^{2+}
(*iv*) Fe

16 Deuterium is an isotope of hydrogen with an atomic mass of 2. Which of the following properties would be affected by an increase in the proportion of deuterium in a given mixture of the two isotopes?

(*i*) the rate of diffusion of the mixture
(*ii*) the heat of combustion per mole of the mixture
(*iii*) the temperature at which the hydrogen could be liquefied at a given pressure
(*iv*) the pressure of the gaseous mixture at standard temperature and pressure

17 Which of the following radioactive emissions travel at the velocity of light?

(*i*) positrons
(*ii*) α particles
(*iii*) β^- particles
(*iv*) γ rays

18 In a complete emission spectrum of the hydrogen atom, in which region(s) will lines appear?

(*i*) the infra-red
(*ii*) the visible
(*iii*) the ultra-violet
(*iv*) the X-ray

Matching pairs questions

Questions 19–22

For each of the elements in questions 19–22, choose from the list A–E the electron configuration which fits the description given.

A $1s^2 2s^2 p^3$
B $1s^2 2s^2 p^6 3s^2 p^6$
C $1s^2 2s^2 p^6 3s^2 p^6 d^6 4s^2$
D $1s^2 2s^2 p^6 3s^2 p^6 d^{10} 4s^2 p^2$
E $1s^2 2s^2 p^6 3s^2 p^6 d^{10} 4s^2 p^4$

19 a noble gas

20 an element with principal oxidation states of +2 and +4

21 an element with principal oxidation states of +2 and +3

22 an element which is isoelectronic with PH_3

Questions 23–26

For each of the nuclear transformations in questions 23–26 choose from the list A–E the category appropriate to the transformation.

A fusion
B fission
C K-capture
D alpha-decay
E beta-decay

23 $^{235}_{92}U \rightarrow ^{141}_{56}Ba + ^{92}_{36}Kr$

24 $^{234}_{90}Th \rightarrow ^{234}_{91}Pa$

25 $^{1}_{0}n \rightarrow ^{1}_{1}H$

26 $^{40}_{19}K \rightarrow ^{40}_{18}Ar$

Assertion-reason questions

	ASSERTION		REASON
27	Fluorine has a lower electron affinity than chlorine	*because*	the chlorine atom needs only one electron to reach a noble gas configuration.
28	A beam of electrons can be diffracted by a nickel crystal	*because*	electrons have wave properties.
29	The second ionization energy of lithium is greater than the second ionization energy of helium	*because*	the Li^+ ion has a noble gas configuration.
30	A beam of silver atoms may be split by a magnetic field	*because*	a magnetic field can cause electron transitions between the 5*d* and 6*s* orbitals.

2 Structure and Bonding

Multiple choice questions

1 Which one of the following most appropriately describes the forces of attraction between molecules in solid rhombic sulphur?

A ionic bonding forces
B covalent bonding forces
C co-ordinate bonding forces
D van de Waals forces
E forces resulting from delocalised electrons

2 Which one of the following salts has the greatest degree of ionic character?

A $BeBr_2$
B KBr
C LiBr
D $MgBr_2$
E $MgCl_2$

3 Which one of the following molecular chlorides would *not*, in the vapour state, have a permanent dipole?

A HCl
B ICl
C SCl_2
D $HgCl_2$
E PCl_3

4 Which one of the following is true about the structure of the carbon dioxide molecule?

A The bond angle is 180°; it has polar bonds and a dipole moment.
B The bond angle is 109°28′; it has polar bonds and a dipole moment.
C The bond angle is 180°; it has polar bonds but no dipole moment.
D The bond angle is 109°28′; it has non-polar bonds and no dipole moment.
E The bond angle is 180°; it has non-polar bonds and no dipole moment.

5 Considering the electron pairs surrounding the central (underlined) atom for which one of the following species is the ratio of lone pairs to bonded pairs greatest?

A $H_2\underline{O}$
B $H_3\underline{O}^+$
C $\underline{N}H_3$
D $\underline{N}H_4^+$
E $\underline{Xe}F_4$

6 Which one of the following species does *not* form co-ordinate bonds with transition metal ions?

A CN^-
B NH_3
C H_2O
D HF
E F^-

7 Which one of the following correctly describes the F—P—F bond angles in a molecule of gaseous phosphorus pentafluoride, PF_5?

A 72° only
B 90° and 120° only
C 90° and 180° only
D 90°, 120° and 180°
E 120° and 180° only

8 Which one of the following undergoes reaction with fluorine?

A KF
B CaF_2
C CF_4
D SF_6
E IF_5

9 Which one of the following in the solid state has a crystal structure which does *not* contain discrete molecules?

A phosphorus pentoxide
B carbon dioxide
C silicon dioxide
D monoclinic sulphur
E white phosphorus

10 Which one of the following data could *not* be used to indicate the presence of hydrogen bonding in a system?

A critical temperature
B enthalpy of vaporisation
C molecular weight
D molar volume
E boiling point

Multiple completion questions

11 Which of the following molecules is/are isoelectronic with methane, CH_4?

(*i*) NH_3
(*ii*) C_2H_4
(*iii*) HF
(*iv*) CCl_4

12 Which of the following is/are linear molecules?

(*i*) NO
(*ii*) H_2O_2
(*iii*) NO_2
(*iv*) $BeCl_2$

13 In which of the following does hydrogen bonding occur?

(*i*) H_2O_2, hydrogen peroxide
(*ii*) C_2H_5OH, ethanol
(*iii*) HF, hydrogen fluoride
(*iv*) NaH, sodium hydride

14 In which of the following species does *each* of the constituent atoms achieve a noble gas configuration?

(*i*) NO
(*ii*) BF_3
(*iii*) PF_5
(*iv*) ClO^-

15 If the structure of a species is best respresented by a resonance hybrid which of the following must be correct?

(*i*) The species is actually an equilibrium mixture of two or more contributing forms.
(*ii*) The species contains delocalised electrons.
(*iii*) The species is uncharged.
(*iv*) The species is actually more stable than any of its contributing forms.

16 On the scale Fluorine = 4.0, elements X and Y have electronegativities of 0.9 and 1.1 respectively. Which of the following properties would be expected from a compound of these two elements?

(*i*) conduction of electricity in the solid state
(*ii*) conduction of electricity in the liquid state
(*iii*) existence as a solid at room temperature and pressure
(*iv*) existence as a gas at room temperature and pressure

Matching pairs questions

Questions 17–21

Choose from the list A–E the geometry which correctly describes the arrangement of atoms around the underlined atom in each of the following compounds.

A linear
B angular
C trigonal planar
D trigonal pyramidal
E tetrahedral

17 $\underline{P}Cl_3$

18 $H\underline{C}{\equiv}CH$

19 $\underline{B}F_3$

20 $BF_3.\underline{N}H_3$

21 $\underline{S}O_2$

Questions 22–25

For each of the phenomena in questions 22–25, choose from the list A–E the conceptual model most useful in its interpretation.

A noble gas electron configuration
B co-ordinate covalent bonding
C hydrogen bonding
D delocalisation of electrons
E hybridisation of atomic orbitals

22 Hydrogen fluoride has a higher boiling point than hydrogen chloride.

23 Graphite conducts electricity.

24 In carbon tetrachloride the chlorine atoms are arranged tetrahedrally around the carbon atom.

25 Copper(II) ions are hydrated in aqueous solution.

Assertion-reason questions

	ASSERTION		REASON
26	The compound $CH_3CH_2CH_2OH$ has a higher boiling point than the compound CH_3CH_2SH	*because*	the compound $CH_3CH_2CH_2OH$ has the higher molecular mass.
27	Hydrochloric acid is a stronger acid than hydrofluoric acid in aqueous solution	*because*	chlorine is more electronegative than fluorine.
28	Sodium ions are not hydrated in aqueous solution.	*because*	the sodium ion has a noble gas electron configuration.
29	$SiCl_4$ reacts with cold water whereas CCl_4 does not	*because*	more electron pairs can be accommodated around the silicon atom than around the carbon atom.
30	Solid aluminium is a good conductor of heat	*because*	delocalised electrons are free to move within solid aluminium.

3 The Gaseous State

Multiple choice questions

1 Which one of the following is *not a necessary* assumption of the kinetic theory of gases?

A Atoms or molecules of a gas are in random motion.
B Atoms or molecules of a gas are much smaller than the distances between them.
C Collisions between atoms or molecules of a gas and the containing vessel are perfectly elastic.
D The kinetic energy of the atoms and molecules in a gas increases as the gas temperature is raised.
E In a given gas, all particles have the same kinetic energy at a given temperature.

2 Which one of the following gases has a density of approximately 0.75 g $litre^{-1}$ at s.t.p.?

A helium
B hydrogen
C neon
D ammonia
E oxygen

3 0.65 g of a gaseous hydrocarbon occupies 1.12 litre at 380 mmHg and 273 K. What is the maximum number of carbon atoms in each molecule of the hydrocarbon?

A 1
B 2
C 3
D 4
E 6

4 What is the approximate density (in g $litre^{-1}$) of sulphur dioxide at 546 K and 2 atmospheres pressure?

A 0.75
B 1.5
C 3
D 6
E 12

5 The mean velocity of helium atoms is 1 200 m sec^{-1} at 273 K. What will be the mean velocity (in m sec^{-1}) of sulphur dioxide molecules under these conditions?

A 1 200
B 400
C 300
D 150
E 75

6 Separate samples of nitrogen and carbon dioxide gas are at the same temperature. Which one of the following will have the same value for the molecules of each gas?

A linear velocity
B rate of diffusion
C linear momentum
D potential energy
E translational kinetic energy

7 An equimolar mixture of hydrogen and oxygen is allowed to diffuse through a small hole. What will be the initial *mass* ratio of hydrogen to oxygen in the diffusing gas?

A 4:1
B 1:4
C 1:2
D 1:8
E 1:16

8 An equimolar mixture of hydrogen and oxygen is allowed to diffuse through a small hole. What will be the initial *molar* ratio of hydrogen to oxygen in the diffusing gas?

A 1:2
B 2:1
C 1:4
D 4:1
E 1:1

9 For which one of the following real gases would the gas constant, *R*, be expected to have the lowest value?

A helium
B hydrogen
C neon
D nitrogen
E ammonia

10 A closed vessel containing gaseous nitrogen was heated from 50°C to 500°C. Which one of the following factors would be *least* affected by this change?

A the average velocity of the molecules
B the average kinetic energy of the molecules
C the range of velocities possessed by individual molecules
D the frequency of intermolecular collisions
E the average momentum of the molecules

11 The value of γ, the ratio of the specific heats at constant pressure and constant volume for three gases are:

neon	1.67
helium	1.67
carbon dioxide	1.29

Why is the value for carbon dioxide different from that for neon and helium?

A Carbon dioxide is a compound whereas neon and helium are elements.
B Carbon dioxide is a reactive gas whereas neon and helium are inert.
C Carbon dioxide has a greater density than either neon or helium.
D Carbon dioxide molecules can rotate and vibrate whereas those of helium and neon cannot.
E Helium and neon approximate more closely to ideal gas behaviour than does carbon dioxide.

12 A sample of hydrogen is collected in a graduated tube over water at 25°C. The column of water remaining in the tube after collection is h mm above the level of water in the surrounding container.

If the atmospheric pressure is P mmHg, the saturated vapour pressure of water p mmHg and the density of mercury 13.6 g cm^{-3} what is the partial pressure of the collected hydrogen?

A $P - p + \frac{h}{13.6}$

B $P - p + 13.6h$

C $P - p - \frac{h}{13.6}$

D $P - p - 13.6h$

E $P + p + \frac{h}{13.6}$

13 A sealed container holds ice, water and water vapour in equilibrium. Which one of the following statements *must* be true?

A The temperature of the container is 0°C if the external pressure is 760 mmHg.
B The average kinetic energy of the molecules in the vapour state is less than the average kinetic energy of the molecules in the solid state.
C In order to get from the solid to the vapour state molecules must pass through the liquid state.
D The average potential energy of molecules in the vapour state is higher than the average potential energy of molecules in the solid state.
E If the pressure within the container is greater than 760 mmHg the temperature within it must be less than 0°C.

14 Three identical containers hold 4 g hydrogen, 4 g oxygen and 4 g of carbon dioxide respectively. Each gas is maintained at the same pressure as the others. Which one of the following represents the relationship between the temperatures (T) of the gases?

A $T_{H_2} = T_{O_2} = T_{CO_2}$
B $T_{H_2} > T_{O_2} > T_{CO_2}$
C $T_{CO_2} > T_{O_2} > T_{H_2}$
D $T_{CO_2} > T_{H_2} > T_{O_2}$
E $T_{O_2} > T_{CO_2} > T_{H_2}$

15 The molecular weights of some liquids may be determined by measurement of their vapour densities. Which one of the following measurements would *not* be essential to such determinations?

A the mass of liquid used
B the boiling point of the liquid
C the temperature at which the vapour was collected
D the pressure at which the vapour was collected
E the volume of the vapour collected

Multiple completion questions

16 Which of the following statements is/are correct?

(*i*) For each gas there is a temperature above which it cannot be liquefied.
(*ii*) Any gas can be condensed to a liquid by pressure alone.
(*iii*) For each substance there is a temperature above which it cannot exist as a liquid whatever the pressure.
(*iv*) A liquid and its vapour are in equilibrium only at the boiling point.

17 Which of the following statements is/are correct?

(*i*) The variation of PV with pressure is governed by the temperature for a given gas.
(*ii*) For an ideal gas a plot of PV against pressure is a straight line with a slope of 45°.
(*iii*) For all real gases a plot of PV against pressure is a straight line with a slope of less than 45°.
(*iv*) Boyle's Law is obeyed only when PV is independent of pressure.

18 If a gas obeys the equation $PV = nRT$ which of the following statements *must* be true for a given mass of that gas?

(*i*) The volume will be inversely proportional to the pressure at a given temperature.
(*ii*) The molecules of the gas have zero volume.
(*iii*) Collisions involving the gas molecules are perfectly elastic.
(*iv*) There is no transfer of kinetic energy during molecular collisions.

19 Which of the following will increase the total number of collisions per second between neon atoms in a given mass of neon?

(*i*) raising the temperature at constant pressure
(*ii*) raising the temperature at constant volume
(*iii*) adding argon at constant temperature and pressure
(*iv*) raising the pressure at constant temperature

20 Which of the following help(s) to explain the difference in properties of a given substance in its liquid and gaseous states?

(*i*) Particles of the substance are closer together in the liquid than in the gaseous state.
(*ii*) The volume of a given mass of gas is governed only by the size of the containing vessel.
(*iii*) Energy is required to convert a liquid to a gas at the boiling point.
(*iv*) A liquid begins to evaporate when the average kinetic energy of the particles is sufficient to enable them to leave the liquid phase.

21 Which of the following *must* increase the net rate of evaporation of a liquid?

(*i*) raising the temperature of the liquid
(*ii*) dissolving an unreactive solid in the liquid
(*iii*) lowering the external atmospheric pressure
(*iv*) adding another inert liquid to it

22 Which of the following statements is true *only* of ideal gases?

(*i*) Under given conditions, equal volumes contain the same number of molecules.
(*ii*) The average kinetic energy of the gas molecules is directly proportional to the absolute temperature.
(*iii*) The total pressure of a mixture of gases is the sum of the partial pressures of the components of the mixture.
(*iv*) The distance between the gas molecules is much greater than the size of the molecules themselves.

Matching pairs questions

Questions 23–30 refer to two identical closed containers P and Q which are at the same temperature. Container P contains 1 mole of helium. Contain Q contains 2 moles of hydrogen. Choose from the list A–E the responses which correctly answer the following questions.

A 1:1
B $1:2(\sqrt{2})$
C $1:\sqrt[3]{2}$
D 1:2
E 1:3

23 What is the ratio of the pressures in P and Q?

24 What is the ratio of the densities of P and Q?

25 What is the ratio of the average distance between particles in P and Q?

26 What is the ratio of the average kinetic energies of the particles in P and Q?

27 What is the ratio of the total kinetic energy of the particles in P and Q?

28 What is the ratio of the degrees of freedom available to particles in P to that available to particles in Q?

29 What is the ratio of the collisions (sec^{-1} cm^{-2}) on the walls of P to those on the walls of Q?

30 What is the ratio of the numbers of nuclear protons in P to that in Q?

4 Energetics

Multiple choice questions

1 Which one of the following data *cannot* be obtained directly by experiment?

A the first ionization energy of magnesium
B the enthalpy of sublimation of magnesium
C the enthalpy of dissociation of chlorine
D the heat of formation of magnesium chloride
E the lattice energy of magnesium chloride

2 Which one of the following best accounts for the fact that aluminium(III) chloride is a readily obtained substance but magnesium(III) chloride is unknown?

A the relative sizes of the Mg^{3+} and Al^{3+} ions
B the relative enthalpies of sublimation of aluminium and magnesium
C the relative ionization energies needed to produce the ions Mg^{3+} and Al^{3+}
D the relative lattice energies of $MgCl_3$(s) and $AlCl_3$(s)
E the ability of aluminium chloride to exist as Al_2Cl_6

3 The first three ionization energies of an element X are 860, 1 700 and 15 000 kJ mol^{-1}. Which one of the following ions is most likely to be present in a compound formed between X and fluorine gas?

A X^+
B X^{2+}
C X^{3+}
D X^-
E X^{2-}

4 Given that

$$C(s) + O_2(g) \rightarrow CO_2(g);\ \Delta H = -a \text{ joules}$$

and $$2CO(g) + O_2(g) \rightarrow 2CO_2(g);\ \Delta H = -b \text{ joules}$$

which one of the following correctly represents ΔH for the heat of formation of carbon monoxide?

A $2a-b$

B $\dfrac{b-2a}{2}$

C $b-2a$

D $\dfrac{2a-b}{2}$

E $\dfrac{b-a}{2}$

5 If the standard enthalpy (heat) of formation of water is -285 kJ mol^{-1}, what is the standard enthalpy of combustion of hydrogen in kJ mol^{-1}?

A -570
B -285
C -142.5
D 0
E $+285$

6 Given that

$$Ce^{4+}(aq) + e^{-} \rightarrow Ce^{3+}(aq);\ E^{\ominus} = +1.61v$$
$$Fe^{3+}(aq) + e^{-} \rightarrow Fe^{2+}(aq);\ E^{\ominus} = +0.77v$$

what is the approximate e.m.f. in volts of the cell represented by the equation

$$Ce^{4+}(aq) + Fe^{2+}(aq) = Ce^{3+}(aq) + Fe^{3+}(aq)?$$

A -2.38
B -0.84
C $+0.84$
D $+0.07$
E $+2.38$

7 Which one of the following properties increases in the order Li, Na, K, Rb, Cs?

A first ionization energy of the metal atom
B hydration energy of the cation
C atomic radius of the metal atom
D melting point of the element
E boiling point of the element

8 The lattice energy of potassium iodide is $-645\,kJ\,mol^{-1}$. The enthalpy of solution of potassium iodide in water is $+22\,kJ\,mol^{-1}$. What is the hydration energy of potassium iodide in $kJ\,mol^{-1}$?

A +667
B +623
C +333.5
D −623
E −667

9 Given that

$$CH_4(g) = C(g) + 4H(g);\ \Delta H = 1\,648\,kJ\,mol^{-1}$$
$$C_2H_6(g) = 2C(g) + 6H(g);\ \Delta H = 2\,810\,kJ\,mol^{-1}$$

what is the value, in $kJ\,mol^{-1}$, of energy required to break up the C—C bond?

A −338
B −1 162
C +332
D +338
E +1 405

10 Which one of the following does *not* vary periodically with atomic number?

A atomic mass
B first ionization energy
C enthalpy of vaporisation
D atomic volume
E atomic radius

Multiple completion questions

11 Which of the following properties contributes to the enthalpy change in the conversion

$$\tfrac{1}{2}X_2(g) \rightarrow X^+(aq)?$$

(*i*) hydration energy of X^+
(*ii*) first ionization energy of X
(*iii*) bond energy X—X
(*iv*) electron affinity of X

12 Which of the following increase(s) in the order HCl, HBr, HI?

(*i*) standard enthalpy of formation (i.e. more heat evolved per mole)
(*ii*) acid strength in aqueous solution
(*iii*) thermal stability
(*iv*) ease of oxidation

13 Which of the following thermodynamic characteristics of the forward reaction in an equilibrium could be altered by the addition of a suitable catalyst?

(*i*) the enthalpy change
(*ii*) the free energy change
(*iii*) the entropy change
(*iv*) the activation energy

14 Which of the following correctly describes the relationship between the enthalpy at constant volume (ΔU) and that at constant pressure (ΔH) for the reaction

$$CH_4(g)+2O_2(g) \rightarrow CO_2(g)+2H_2O(g)?$$

(*i*) ΔH is less than ΔU
(*ii*) ΔH is the same as ΔU
(*iii*) ΔH is greater than ΔU
(*iv*) both ΔH and ΔU will be negative as heat is evolved

15 A substance X reacts separately with substances P and Q to form products PX and QX respectively. When 1 mole of X reacts with a mixture containing 1 mole of each of P and Q at low temperatures the main product is PX. At higher temperatures, the same reaction mixture produces mainly QX. Which of the following statements is/are correct?

(*i*) The produce PX has greater thermodynamic stability than QX.
(*ii*) The activation energy for the reaction $P+X \rightarrow PX$ is less than that for the reaction $Q+X \rightarrow QX$.
(*iii*) The activation energy for the reaction $Q+X \rightarrow QX$ is less than that for the reaction $P+X \rightarrow PX$.
(*iv*) The product QX has a greater thermodynamic stability than PX.

16 Which of the following properties of the noble gases increase(s) with increasing atomic number?

(*i*) atomic volume
(*ii*) molar enthalpy of vaporisation
(*iii*) boiling point
(*iv*) first ionization energy

Matching pairs questions

Questions 17–20

Choose, from the list A–E, the most appropriate description of the energy change involved in each of the reactions specified in questions 17–20.

A oxidation potential
B enthalpy of ionization
C enthalpy of solution
D enthalpy of formation
E enthalpy of neutralisation

17 $H(g) \rightarrow H^+(g) + e^-$

18 $\frac{1}{2}H_2(aq) \rightarrow H^+(aq) + e^-$

19 $C(s) + \frac{1}{2}O_2(g) \rightarrow CO(g)$

20 $NH_2^-(NH_3) + NH_4^+(NH_3) \rightarrow 2NH_3$

Questions 21–24

Choose, from the list A–E, the value which correctly corresponds to the energy change involved in each of the reactions specified in questions 21–24. (Heat evolved is given a negative sign.)

A −3 kJ
B −60 kJ
C −300 kJ
D −3 400 kJ
E +200 kJ

21 $4Al(s) + 3O_2(g) \rightarrow 2Al_2O_3(s)$

22 $H^+(aq) + OH^-(aq) \rightarrow H_2O(l)$

23 $N_2(g) + O_2(g) \rightarrow 2NO(g)$

24 $(CH_3)_2C{=}O + CHCl_3 \rightarrow (CH_3)_2C{=}O \cdots HCCl_3$

Assertion-reason questions

	ASSERTION		REASON
25	Calcium forms a chloride of formula $CaCl_2$	*because*	$CaCl_2$ is thermodynamically more stable than either CaCl, $CaCl_3$ or a mixture of calcium and chlorine.
26	The standard enthalpy of formation of $SrCl_2$ is greater than that of $CaCl_2$	*because*	the first ionization energy of Sr is greater than that of Ca.
27	Radium undergoes spontaneous radioactive decay	*because*	radium has the highest first ionization energy of any Group II element.
28	The first ionization energy of the atom increases in the order Na, Mg, Al	*because*	the atomic number also increases in this order.
29	Endothermic reactions proceed more rapidly as the temperature is raised	*because*	the heat increases the kinetic energy of the reactant molecules.
30	Compounds of Cu(I) are unknown	*because*	Cu(I), on contact with water, disproportionates to Cu(II) and Cu(O).

5 Kinetics

Multiple choice questions

1 Which one of the following statements correctly describes the term "activation energy"?

A the energy which must be applied to a reaction before it will take place
B the minimum energy required for molecules to react successfully
C the potential energy of the activated complex
D the minimum bond energy among reacting species
E the energy difference between reactants and products

2 Which one of the following is a *necessary* condition for a reaction to be second order with respect to a reactant, X?

A There is at least one reactant in addition to X.
B The reaction involves two steps.
C The rate determining step involves two molecules of X.
D The reaction mechanism involves at least two molecules of X.
E Only two molecular species take part in the reaction.

3 For a reaction whose rate expression is

$$\text{rate}_1 = k[P][Q]^2$$

increasing the concentration of P to $3P$ while that of Q remains unchanged will yield a new rate, rate_2, which is related to rate_1 by a factor of

A 1
B 2
C 3
D 6
E 9

4 For an overall reaction $2V + W = Y + Z$, a proposed mechanism is

$$V + W \rightarrow VW \quad \text{fast}$$
$$VW \rightarrow X + Y \quad \text{slow}$$
$$X + V \rightarrow Z \quad \text{fast}$$

Which one of the following experimental rate laws supports this proposed mechanism?

A rate = $k[V]^2[W]$
B rate = $k[V][W]$
C rate = $k[V]^2[W][X]$
D rate = $\dfrac{k[Y][Z]}{[V]^2[W]}$
E rate = $k[V]^2[W]^2$

5 Solutions containing Ce^{4+}(aq) and Tl^{+}(aq) ions react according to the equation

$$2Ce^{4+}(aq) + Tl^{+}(aq) = 2Ce^{3+}(aq) + Tl^{3+}(aq)$$

On mixing the aqueous solutions the reaction is slow but on the addition of Mn^{2+}(aq) it occurs much more rapidly. Which one of the following statements *must* be true?

A The activation energy for the interaction of Ce^{4+}(aq) and Tl^{+}(aq) is greater than that for the interaction of Ce^{4+}(aq) and Mn^{2+}(aq).
B The activation energy for the interaction of Tl^{+}(aq) and Mn^{2+}(aq) is less than that for the interaction of Tl^{+}(aq) and Ce^{4+}(aq).
C The presence of Tl^{+}(aq) ions inhibits the reaction between Ce^{4+}(aq) and Mn^{2+}(aq).
D During the catalysed reaction, Ce^{4+}(aq) oxidises Mn^{2+}(aq) to Mn^{4+}(aq) which Tl^{+}(aq) reduces back to Mn^{2+}(aq).
E The rate law for the reaction is rate = $k[Ce^{4+}(aq)]^2[Tl^{+}(aq)]$.

6 In the reaction

$$2NO(g) + 2H_2(g) = N_2(g) + 2H_2O(g)$$

carried out at 1 000 K, initial partial pressures of 200 mmHg for hydrogen and 150 mmHg for nitric oxide resulted in a fall in pressure during the first 100 seconds of the reaction of 12.5 mmHg. If the initial partial pressure of each of the reactants is doubled what will be the loss in pressure (in mmHg) during the same period of time?

A 25
B 50
C 100
D 200
E unobtainable from the given data

7 The rate expression for a reaction between two substances P and Q in aqueous solution to give a gaseous product is

$$\text{rate} = k[\text{P}]^2[\text{Q}]$$

At given concentrations of P and Q, 90 cm^3 of gas were produced in the first 100 seconds of reaction. If the initial concentration of P is doubled and the initial concentration of Q halved, what will be the volume of gas (in cm^3) produced under the same external conditions in the same period of time?

A 45
B 90
C 180
D 270
E unobtainable from the data given

8 Which one of the following statements is true concerning the reaction

$${}^{210}_{84}\text{Po} = {}^{206}_{82}\text{Pb} + {}^{4}_{2}\text{He} \quad ?$$

A The reaction is first order with respect to polonium, Po.
B It is a zero order reaction.
C The rate of reaction is reduced by combining the elemental polonium with other elements.
D The rate of reaction may be varied by altering the state of sub-division of the reactants.
E The rate of reaction may be reduced by enclosing the reactant in a lead container.

9 Which one of the following statements about the reaction between gaseous hydrogen and chlorine is *false*?

A It may be initiated by ultra-violet light.
B In ultra-violet light a free radical reaction takes place.
C It may be initiated by blue but not by red light.
D It takes place more slowly in diffused light than in ultra-violet light.
E Photons atomise the hydrogen and chlorine molecules which then interact.

10 In a chemical reaction $P+Q=R+S$, the rate of the reaction is described by the equation

$$\text{rate} = k[P]^{0.5}[Q]$$

Which one of the following statements is *incorrect*?

A The rate of the reaction is doubled if the concentration of Q is doubled.
B The overall order of the reaction is 1.5.
C The reaction is first order with respect to Q.
D The order of the reaction is 0.5 with respect to P.
E The rate of the reaction is halved if the concentration of P is halved.

11 The silver ion in 0.1M aqueous solution will react with each of (*i*) 0.1M Cl^- (aq) (*ii*) Cu(s) and (*iii*) 0.1M NH_3(aq). What would be the relative rates of reaction of the silver ion with these species under these conditions?

A (*i*) > (*ii*) > (*iii*)
B (*i*) > (*iii*) > (*ii*)
C (*i*) = (*iii*) > (*ii*)
D (*iii*) > (*i*) > (*ii*)
E (*iii*) > (*ii*) > (*i*)

Multiple completion questions

12 Which of the following is/are affected by the addition of a catalyst to a reaction?

(*i*) the activation energy
(*ii*) the enthalpy change
(*iii*) the reaction mechanism
(*iv*) the position of equilibrium

13 The rate expresson for a given reaction is

$$\frac{-\mathrm{d}[A]}{\mathrm{d}t} = k[A]$$

Which of the following statements about the reaction must be true?

(*i*) It is a first order reaction.
(*ii*) A plot of log $[A]$ against t at a given temperature is a straight line.
(*iii*) The reaction could be a nuclear disintegration.
(*iv*) The rate of the reaction is independent of the temperature.

14 If the order of a reaction with respect to a reactant, R, is zero, this means that

(*i*) the reaction rate is independent of temperature
(*ii*) the reaction does not occur to any appreciable extent
(*iii*) the rate law cannot be experimentally determined
(*iv*) the reaction rate is independent of the concentration of R

15 In a reaction between substances A and B some collisions between molecules of A and of B do not lead to a chemical reaction. The reason for this may be that

(*i*) the system is already at equilibrium
(*ii*) the molecules of A and B have insufficient energy
(*iii*) the distribution of energy between A and B is uneven
(*iv*) the orientation of A and B at collision is inappropriate

16 Which of the following is/are an *essential* characteristic(s) of a catalyst?

(*i*) It must be in the same physical state as the reactants.
(*ii*) It must remain chemically unchanged at the end of the reaction.
(*iii*) It must possess a large surface area.
(*iv*) It must provide an alternative reaction mechanism involving a lower activation energy.

17 Which of the following statements concerning a photochemical reaction is/are *necessarily* true?

(*i*) It is initiated only by photons of a specific energy.
(*ii*) Once the reaction is initiated it will proceed explosively.
(*iii*) The rate of the reaction is directly proportional to the intensity of the light employed.
(*iv*) There is a threshold energy below which reaction is not initiated.

Matching pairs questions

Questions 18 to 22 refer to the diagram below which shows the energy profiles for a reaction with, and without, a catalyst present. Choose from the keys A–E the energy changes which correctly answer each of questions 18 to 22.

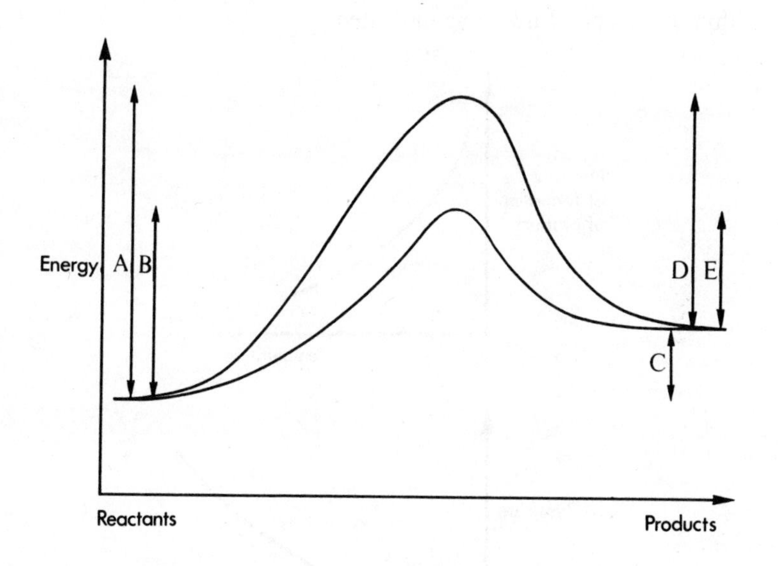

18 Which is the activation energy of the uncatalysed forward reaction?

19 Which is the enthalpy change for the uncatalysed reverse reaction?

20 Which is the activation energy for the catalysed reverse reaction?

21 Which is the activation energy of the reaction which will proceed most rapidly at high temperatures?

22 Which is the activation energy of the reaction which is least likely to occur at low temperatures?

Questions 23–26

Each of the diagrams A to E below represents a possible dependence on temperature of the rate at which product(s) are formed in a chemical reaction. Choose from the diagrams A–E the diagram which most closely corresponds to each of the reactions specified in questions 23 to 26. In each case consider the reaction only within the temperature range indicated.

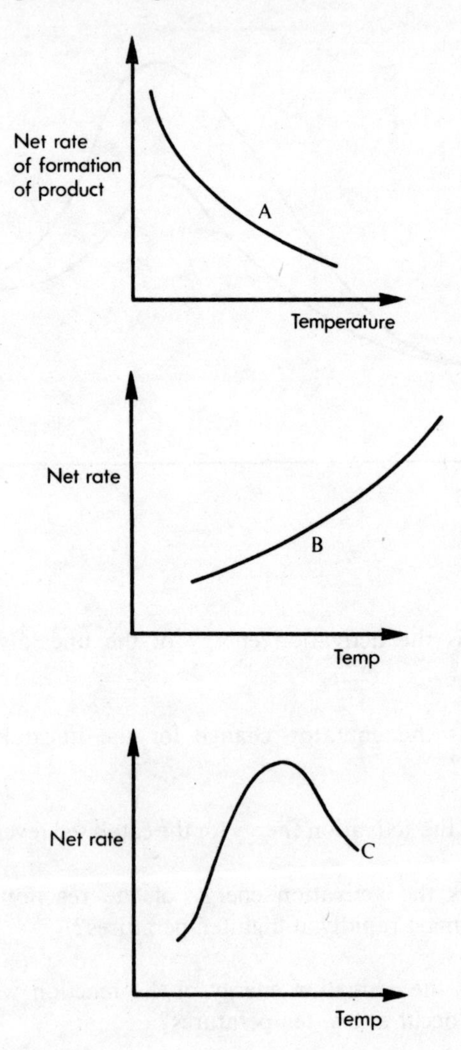

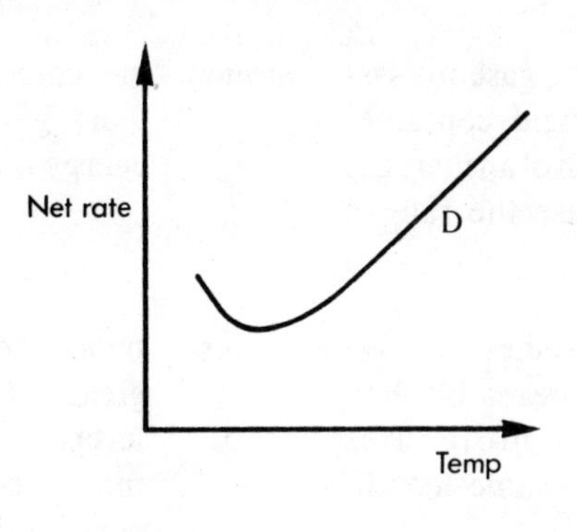

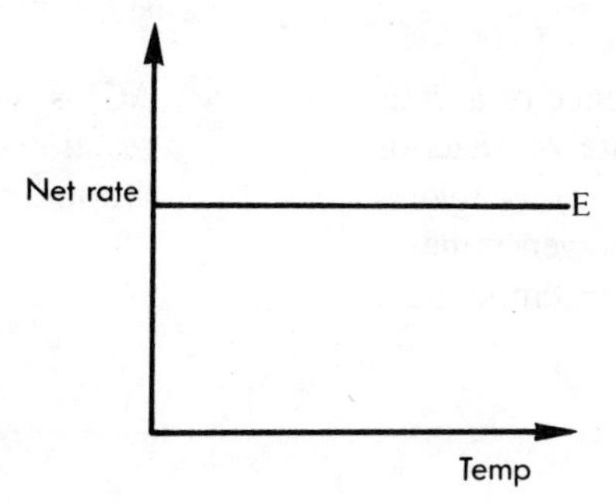

23 $^{226}_{88}Ra = ^{222}_{86}Rn + ^{4}_{2}He$; 20°C–300°C

24 $PCl_3(g) + Cl_2(g) = PCl_5(g)$; 200°C–300°C

25 $C_{12}H_{22}O_{11} + H_2O \overset{\text{invertase}}{=} C_6H_{12}O_6 + C_6H_{12}O_6$; 20°C–80°C

26 $Zn(s) + 2HCl(aq) = ZnCl_2(aq) + H_2(g)$; 20°C–50°C

Assertion-reason questions

	ASSERTION		REASON
27	An inhibitor reduces the rate of a chemical reaction	*because*	inhibitors reduce the fraction of reactant molecules with a given kinetic energy.

28	In a reacting gaseous system in a rigid container the addition of an inert gas will decrease the rate of reaction	*because*	in collisions involving inert gas molecules no energy is exchanged.
29	Polar molecules dissolve in water more readily than non-polar molecules under the same conditions	*because*	polar molecules have a greater range of energy levels than non-polar molecules under the same conditions.
30	In the absence of a catalyst the rate of reaction between gaseous hydrogen and oxygen is negligible at room temperature	*because*	ΔG is positive for this reaction under these conditions.

6 Equilibria I

Multiple choice questions

1 An acid HX has a dissociation constant of 1.6×10^{-4} mol litre^{-1} at 25°C. What is the hydrogen ion concentration in g ion litre^{-1} of a 0.1M aqueous solution of HX at this temperature?

A 1.2×10^{-2}
B 3.2×10^{-2}
C 4×10^{-3}
D 1.6×10^{-6}
E 2.6×10^{-8}

2 If K_a for acetic acid and K_b for aqueous ammonia are each 10^{-5} mol litre^{-1} at 25°C, which of the following will have a pH of about 11 at this temperature?

A a solution of 0.1M ammonia and 0.1M ammonium acetate
B a mixture of equal volumes of M sodium acetate and M acetic acid
C a mixture of equal volumes of M ammonia and M acetic acid
D a molar solution of ammonia
E a decimolar solution of ammonium acetate

3 In which one of the following reactions will an increase in total pressure at constant temperature favour the formation of products?

A $CaCO_3(s) \rightarrow CaO(s) + CO_2(g)$
B $2C(s) + O_2(g) \rightarrow 2CO(g)$
C $2NO(g) + O_2(g) \rightarrow 2NO_2(g)$
D $H_2(g) + Br_2(g) \rightarrow 2HBr(g)$
E $2NH_3(g) \rightarrow 3H_2(g) + N_2(g)$

4 The pH ranges of some indicators are as follows:

(*i*) methyl violet 0.5 – 1.5
(*ii*) methyl yellow 2.9 – 4.0
(*iii*) methyl orange 3.1 – 4.4
(*iv*) methyl red 4.2 – 6.3
(*v*) phenolphthalein 8.0 – 9.6

Which of the above indicators could be used to standardise a solution of hydrocyanic acid ($K_a = 4 \times 10^{-10}$ mol litre^{-1} at 25°C) against potassium hydroxide solution?

A (*i*) only
B (*v*) only
C (*ii*) and (*iii*) only
D (*i*), (*ii*), (*iii*) and (*iv*) only
E (*iv*) and (*v*) only

5 For the liquid phase equilibrium, $V + W \rightleftharpoons X + Y$, the equilibrium constant at 20°C is 81. If 1 mole of each of V and W are mixed and allowed to attain equilibrium at 20°C, what fraction of a mole of X is obtained?

A 0.1
B 0.3
C 0.4
D 0.5
E 0.9

6 The equilibrium constant for the following equilibrium is 49×10^{-5} mol litre^{-1} at 25°C.

$$CH_3NH_2 + H_2O \rightleftharpoons CH_3NH_3^+ + OH^-$$

What is the approximate concentration of hydroxide ions in mol litre^{-1} in a 0.1M solution of methylamine at 25°C?

A 7×10^{-3}
B 49×10^{-3}
C 2.2×10^{-2}
D 7×10^{-4}
E 4.9×10^{-3}

7 Which one of the following aqueous solutions has the largest concentration of hydroxide ions?

A 0.5M Na_2CO_3
B 0.05M $Ba(OH)_2$
C 0.05M KOH
D M $NaHCO_3$
E M $PO(OH)_3$

8 Which one of the following statements is correct?

A Maleic acid and fumaric acid have the same melting point.
B Fumaric acid on heating forms maleic anhydride.
C Fumaric acid exhibits optical isomerism whereas maleic acid does not.
D Maleic and fumaric acids have the same solubility in water.
E The first acid dissociation constant for maleic acid is greater than that for fumaric acid.

9 The acid HY is found to be 1% ionized in 0.01M aqueous solution. What is the value of pK_a for the acid HY under these conditions?

A 2
B 4
C 5
D 6
E 8

10 What is the approximate ratio of the first acid dissociation constant to the second acid dissociation constant for phosphoric acid, H_3PO_4?

A 10^5
B 10
C 1
D 10^{-1}
E 10^{-5}

11 In the equilibrium $P+Q \rightleftharpoons R+S$, one mole of each of P and Q are mixed. At equilibrium 0.333 mole of P are found to be present. What is the value of the equilibrium constant under these conditions?

A 0.667
B 2.0
C 4.0
D 6.0
E 8.0

12 Which one of the following correctly accounts for the fact that phenolphthalein and not methyl orange is used as an indicator in titrating oxalic acid with sodium hydroxide?

A Sodium hydroxide is a strong base.
B Oxalic acid is a dibasic acid.
C Sodium oxalate solution has a pH greater than 7.
D Oxalic acid is too weak an acid to effect a colour change in methyl orange.
E The end-point with phenolphthalein involves a more readily detected colour change than occurs with methyl orange.

13 Each of the following statements refers to a solution of a strong, monobasic acid. Which one of the statements is correct?

A The apparent degree of ionization is independent of the concentration.
B If a molar solution is diluted, the molar conductance will vary directly with the concentration.
C The pH of a 1.0M solution will be greater than 1.
D The degree of dissociation of the acid will approximate to 1 as it is progressively diluted.
E The degree of dissociation of the acid in molar solution will increase as the temperature is raised.

Multiple completion questions

14 Which of the following changes result(s) in the displacement of the stated equilibrium in favour of the right hand side?

(*i*) $PCl_5(s) \rightleftharpoons PCl_3(l) + Cl_2(g)$	increase in pressure
(*ii*) $2NO_2(g) \rightleftharpoons N_2O_4(g)$	increase in pressure
(*iii*) $CuSO_4 . 5H_2O(s) \rightleftharpoons CuSO_4 . 3H_2O(s) + 2H_2O(g)$	increase in concentration of $H_2O(g)$
(*iv*) $N_2(g) + 3H_2(g) \rightleftharpoons 2NH_3(g)$	decrease in temperature (ΔH is −ve for forward reaction)

15 Henry's Law states that the solubility of a gas in a liquid at a given temperature is directly proportional to the pressure. Which of the following gases would *not*, even approximately, obey this law?

(*i*) NH_3
(*ii*) O_2
(*iii*) Ar
(*iv*) HCl

16 Which of the following statements is/are correct?

(*i*) The solubility of all gases in water decreases as the temperature is raised.
(*ii*) The solubility of a given gas in water is unaffected by the presence of any other gas with which neither it nor the solvent chemically reacts.
(*iii*) The solubility of any gas increases as the pressure is raised at a given temperature.
(*iv*) Any gas can be completely expelled from its solution in a liquid by boiling.

17 Which of the following conditions govern(s) the application of Raoult's Law which relates the relative lowering of the vapour pressure of a solvent to the mole fraction of the solute in the solvent?

(*i*) an involatile solute
(*ii*) a dilute solution
(*iii*) a given temperature
(*iv*) an involatile solvent

Matching pairs questions

Questions 18–21

Choose from the list A–E, the acids which correctly answer each of questions 18–21.

A H_3PO_4
B H_2SO_4
C $HClO_4$
D H_3PO_3
E HNO_3

18 Which of the above acids in 0.1M aqueous solution would have the highest concentration of hydroxonium ions?

19 Which of the above acids in 0.1M aqueous solution would have the highest pH?

20 Which of the above acids never acts as a proton acceptor?

21 25 cm^3 of a 0.1M solution of each of the above acids is treated separately with 75 cm^3 of 0.1M NaOH. In which case does the pH of the resulting solution have its lowest value?

Questions 22–25

For each of the reactions described in questions 22–25 choose, from the list A–E, the formula of the species most likely to be formed.

A $AlCl_3$
B Al_2Cl_6
C $[Al(H_2O)_6]^{3+}$
D $[Al(OH)_6]^{3-}$
E $Al(OH)_3$

22 The reaction of aluminium with excess hot, 50% sulphuric acid.

23 The reaction of aluminium with excess chlorine at 1000°C.

24 The reaction of $[Al(H_2O)_6]^{3+}$ with excess, aqueous ammonia.

25 The reaction of aqueous $[Al(H_2O)_6]^{3+}$ ions with excess, aqueous sodium hydroxide.

Assertion-reason questions

	ASSERTION		REASON
26	When a weak acid is added to a weak alkali, the electrical conductivity of the solution will fall to a minimum before rising again	*because*	the neutralisation of any acid by an alkali in aqueous solution may be represented by the equation $H^+(aq) + OH^-(aq) \rightarrow H_2O(l)$
27	Monochloroacetic acid has a higher K_a than acetic acid at a given temperature	*because*	the substitution of a chlorine atom for a hydrogen atom in the acetic acid molecule increases the molecular weight of the acid.
28	The heats of neutralisation of nitric acid and hydrochloric acid by sodium hydroxide are approximately the same ($-57\,kJ\,mol^{-1}$)	*because*	$-57\,kJ$ is the heat of formation of water.
29	Steam distillation effectively lowers the temperature at which aniline can be distilled	*because*	aniline is denser than water.
30	The equilibrium vapour pressures of liquids decrease with increasing temperature	*because*	the rate at which particles leave a liquid increases as the temperature is raised.

Equilibria II

Multiple choice questions

1 The predicted molecular weight of PCl_5 is 208. If the actual molecular weight of a sample of phosphorus pentachloride in the vapour phase is 156, what is its degree of dissociation?

A 0.75
B 0.66
C 0.33
D 0.25
E 0.125

2 If N_2O_4 (molecular weight = 92) is 50% dissociated into NO_2 at a given temperature what would be the apparent molecular weight?

A 92
B 69
C 62
D 50
E 46

3 Which of the following statements is *incorrect*?

A The osmotic pressure of a dilute solution is the same as the pressure the solute would exert if it existed as a gas at the same temperature and occupied the same volume as the solution.
B Under ideal conditions, the osmotic pressure exerted by a solution at a given temperature is proportional to the concentration of solute.
C For any given dilute solution, the ratio of osmotic pressure to absolute temperature is constant.
D If two solutions X and Y each have an osmotic pressure equal to that of solution Z, the osmotic pressures of X and Y must also be equal.
E Osmotic pressure is the pressure which causes the passage of solvent into a more concentrated solution across a semi-permeable membrane.

4 A 0.01 M solution of sucrose in water freezes at −0.018°C. What would be the freezing point (in °C) of a 0.01 M solution of acetic acid under the same conditions?

A −0.018
B −0.036
C −0.009
D −0.019
E −0.017

5 Which one of the following would give the largest depression in the freezing point when 0.001 mole is dissolved in 100 g of water?

A sodium chloride
B zinc sulphate
C urea
D barium chloride
E sucrose

6 Under which one of the following conditions does the Partition Law, governing the distribution of a solute between two solvents, cease to apply?

A The solvents are miscible in all proportions.
B Each of the solvents is unsaturated with solute.
C The temperature is kept constant.
D The solute is in the same molecular form in both solvents.
E The concentration of solute in one solvent is twice its concentration in the other.

7 When potassium nitrate is heated with concentrated sulphuric acid, nitric acid may be collected as a distillate. Which one of the following explains this fact?

A Sulphuric acid is a stronger acid than nitric acid.
B Nitric acid is more volatile than sulphuric acid.
C Potassium sulphate is more stable to heat than potassium nitrate.
D Nitric acid decomposes more readily than sulphuric acid.
E The dissociation constant of nitric acid is greater than either of the dissociation constants for sulphuric acid.

8 Which one of the following explains why acetic acid is not normally titrated with aqueous ammonia?

A The pH at the equivalence point is less than 7.
B The pH at the equivalence point is 7.
C The pH change near the equivalence point is small.
D The rate of reaction between ammonia and acetic acid is too slow.
E The dissociation constants of acetic acid and aqueous ammonia are almost the same.

9 What is the approximate value of the pH of a solution of 50 cm^3 of M NaOH to which 49 cm^3 of M HCl has been added?

A 14
B 13
C 12
D 11
E 10

10 In a molar solution of potassium phosphate, K_3PO_4, which one of the following concentration relationships is correct?

A $3[K^+(aq)] = [PO_4^{3-}(aq)]$
B $[PO_4^{3-}(aq)] > [HPO_4^{2-}(aq)] > [H_2PO_4^-(aq)]$
C $[H_2PO_4^-(aq)] > [HPO_4^{2-}(aq)] > [PO_4^{3-}(aq)]$
D $[H_3O^+(aq)] > [OH^-(aq)]$
E $[H_3O^+(aq)] = [OH^-(aq)]$

Multiple completion questions

11 Which of the following indicators may be used in the standardisation of potassium hydroxide using potassium hydrogenphthalate? (K_a for phthalic acid is 1.1×10^{-3} mol litre^{-1} at 25°C)

(*i*) methyl yellow (pH range 2.9–4.0)
(*ii*) methyl orange (3.1–4.4)
(*iii*) methyl red (4.2–6.3)
(*iv*) phenolphthalein (8.0–9.6)

12 Which of the following statements about a 0.01M aqueous solution of acetic acid is/are correct? (pK_a for acetic acid at 25°C = 4.8)

(*i*) its pH is 4.8
(*ii*) the concentration of acid is approximately 10^{-2} mol litre^{-1}
(*iii*) the concentration of (hydrated) hydrogen ions is 4.8×10^{-2} mol litre^{-1}
(*iv*) the acid dissociation constant at 25°C is $10^{-4.8}$

13 Which of the following directly govern(s) the saturated vapour pressure of a solution?

(*i*) the nature of the solvent
(*ii*) the concentration of the solution
(*iii*) the temperature of the solution
(*iv*) the vapour density of the solvent

14 Which of the following statements is/are correct?

(*i*) Deliquescence is a relative property, depending upon the actual pressure of water vapour in the atmosphere.
(*ii*) By suitably changing the conditions a given substance may become either efflorescent or deliquescent.
(*iii*) A substance is efflorescent when the vapour pressure of its hydrate exceeds that of the water vapour in the air.
(*iv*) A substance must be deliquescent if its vapour pressure is less than the vapour pressure of the water in surrounding atmosphere.

15 When a mixture of aniline (b.p. = 184°C) and water (b.p. = 100°C) is heated to boiling point, which of the following is true?

(*i*) The boiling point of the mixture is lower than that of either aniline or water.
(*ii*) The distillate obtained by condensing the vapour of the boiling mixture will contain a higher proportion of aniline by weight than the original mixture.
(*iii*) The mixture boils when the sum of the vapour pressures of the aniline and water equals the atmospheric pressure.
(*iv*) The mass ratio of aniline to water in the distillate will be the same as the ratio of the vapour pressure of aniline to that of water at the boiling point of the mixture.

16 Which of the following quantitative relationships concerning a chemical system would indicate that the system *must* be in a state of equilibrium?

(*i*) The free energy change, ΔG, is zero.
(*ii*) The entropy change, ΔS is zero.
(*iii*) The electrode potential, E, is zero.
(*iv*) $\Delta G = \Delta H - T\Delta S$ where T is the absolute temperature.

17 A dilute solution of a solute in water is said to have a freezing point of $-0.043°C$. Which of the following is/are true at $-0.043°C$?

(*i*) The solution freezes.
(*ii*) Crystals of solute begin to appear.
(*iii*) The solution is saturated with solute.
(*iv*) Crystals of ice begin to appear.

18 A dilute solution of a solute in 100 g of water begins to freeze at $-0.10°C$. Which of the following statements is/are true?

(*i*) As the solution is cooled further, more ice will crystallise from the solution.
(*ii*) At a temperature of $-0.40°C$, three quarters of the water will have frozen to form ice.
(*iii*) The temperature of the solution continues to drop from $-0.10°C$ as freezing proceeds.
(*iv*) At a temperature of $-0.40°C$, the solution is three times as concentrated with respect to the solvent as at $-0.10°C$.

19 Which of the following properties of a substance is/are *essential* if it is to be extracted by steam distillation?

(*i*) a low molecular mass
(*ii*) a low vapour pressure
(*iii*) a melting point above room temperature
(*iv*) a low solubility in water

20 Which of the following conditions *must* apply if the molecular weight of a solute is to be determined by the method of measuring the depression of the freezing point?

(*i*) The solute must be capable of vaporisation without decomposition.
(*ii*) The solute must remain entirely in the solution.
(*iii*) The solute and solvent must readily form a range of solid solutions.
(*iv*) The solution must be dilute.

Assertion-reason questions

	ASSERTION		REASON
21	The equilibrium constant for the equilibrium $2Cu^{+}(aq) \rightleftharpoons Cu^{2+}(aq) + Cu(s)$ is less than 1	*because*	the lattice energy of copper metal is greater than the first ionization energy of copper.

22 Cobalt (II) sulphate in aqueous solution turns pink when excess hydrochloric acid is added *because* cobalt (II) chloride is formed under these conditions.

23 At a given pressure sulphur dioxide is less soluble in hot water than in cold *because* the dissolution of sulphur dioxide in water is an endothermic process.

24 When 0.4 g of sodium hydroxide($M = 40$) is dissolved in a litre of water the pH of the solution is 10 *because* the concentration of hydrogen ions in the solution is 10^{-4} mol litre^{-1}.

25 Carbon dioxide is liberated when sodium carbonate is strongly heated with silica *because* carbon dioxide is more volatile than silicon dioxide.

26 A catalyst cannot affect the composition of an equilibrium mixture *because* the composition of an equilibrium mixture depends only upon the temperature.

27 A solid may be completely converted into a liquid without a change in its temperature *because* the average energy of the particles may remain constant during this change.

28 The equilibrium constant for the equilibrium $N_2O_4(g) \rightleftharpoons 2NO_2(g)$ increases with increasing temperature *because* the rate at which $N_2O_4(g)$ molecules dissociate increases with increasing temperature.

29 0.1M nitric acid has a slightly different pH from 0.05 M sulphuric acid *because* sulphuric acid is a stronger acid than nitric acid.

30	The addition of $Cl^-(aq)$ ions to aqueous potassium iodide liberates iodine	*because*	iodine is only slightly soluble in water.

8 The Periodic Table

Multiple choice questions

For questions 1–4 imagine that the Pauli exclusion principle permitted 3 electrons per orbital.

Assume that the masses and charges of the subatomic particles are unaltered and that the only difference in their behaviour is in the formation of electron 'triples'.

1 Which one of the following would correctly describe the element of atomic number 15?

A a strongly electropositive metal
B a strongly electronegative non-metal
C an inert gas
D a non-metal with principal oxidation state +5
E a metal with principal oxidation state +3

2 What would be the stoicheiometric formula of the compound formed between the element of atomic number 2, X, and the element of atomic number 4, Y?

A X_2Y_4
B XY_2
C X_2Y
D XY
E XY_3

3 Which of the first 10 elements would exist as a diatomic gas at room temperature?

A element 1 only
B element 2 only
C elements 1 and 2 only
D elements 2, 4 and 8 only
E elements 2 and 8 only

4 Which one of the following statements concerning the "revised" periodic table is correct?

A It contains more elements than the present periodic table.
B It would contain the same proportion of transition elements as the present table.
C It would not contain paramagnetic elements.
D It would comprise 12 rather than 8 major groups.
E Element of atomic number 7 would be in Group III of the table.

5 For which one of the following is the radius ratio the greatest?

A Li^+/Li
B Be^{2+}/Be
C N^{3-}/N
D O^{2-}/O
E F^-/F

6 If an element has successive ionization energies of 800, 1 600, 13 600, 22 000 and 27 000 kJ mol^{-1}, to which group of the periodic table would you expect it to belong?

A I
B II
C III
D IV
E V

7 In which one of the following cases does the first named element *not contain* 18 more electrons per atom than the second?

A xenon and krypton
B chromium and carbon
C ruthenium and iron
D barium and strontium
E potassium and lithium

8 If a transition metal atom or ion is defined as one containing an incomplete *d* shell of electrons which one of the following would be classified as a transition species?

A Cu^+
B Zn^{2+}
C Ta^{3+}
D Tl^+
E Hg^+

9 In which one of the following groups of the periodic table is the electronegativity difference between the first and last members at a maximum?

A Group IA (Li-Fr)
B Group IIIA (B-Tl)
C Group IVB (C-Pb)
D Group VB (N-Bi)
E Group VIB (O-Po)

10 Which one of the following pairs of elements forms only one binary compound?

A xenon and fluorine
B potassium and nitrogen
C sulphur and chlorine
D iodine and chlorine
E sodium and oxygen

11 Which one of the following elements does not form both ionic and covalent binary compounds?

A hydrogen
B caesium
C aluminium
D oxygen
E fluorine

Multiple completion questions

12 Which of the following statements about the periodic table is/are correct?

(*i*) Most elements within it are metals.
(*ii*) The electronegativity of the elements in any group increases with increasing atomic number.
(*iii*) All naturally occurring elements which are radioactive have an atomic number which is greater than that of bismuth (=83).
(*iv*) The periodicity of properties of the elements is a consequence of the structure of the atom.

13 Which of the following statements about the element francium is/are correct?

(*i*) All its known isotopes are radioactive.
(*ii*) It is an *s* block element.
(*iii*) The ion Fr^+(aq) is colourless.
(*iv*) It occurs in nature as the double salt, $Fr_2CO_3 . FrHCO_3$.

14 It is reported that a relatively stable isotope of element 114 will be synthesised in the not too distant future. Which of the following properties cannot reasonably be predicted for this nuclide?

(*i*) its electron configuration
(*ii*) the range of oxidation states
(*iii*) the physical state of the element at s.t.p.
(*iv*) the crystal structure of the element in the solid state

15 Which of the following statements is/are true?

(*i*) The chemistry of cadmium resembles that of mercury more closely than it does that of zinc.
(*ii*) Copper forms a greater variety of stable complexes than does potassium.
(*iii*) The chemistry of beryllium resembles that of silicon more than it does that of aluminium.
(*iv*) The oxides of iodine are more thermally stable than those of chlorine.

16 Which of the following statements is/are correct?

(*i*) An atom of an element of odd atomic number contains one or more unpaired electrons when in its ground state.
(*ii*) No element of even atomic number contains an atom with an unpaired electron when in its ground state.
(*iii*) All unpaired electrons have the same spin quantum number.
(*iv*) There is experimental evidence for the existence of unpaired electrons.

17 Which of the following is/are isoelectronic with argon?

(*i*) S^{2-}
(*ii*) K^+
(*iii*) Cl^-
(*iv*) Ne

18 Which of the following series of oxides are arranged in order of increasing basicity?

(*i*) Na_2O, CaO, Al_2O_3, I_2O_5
(*ii*) CO_2, SnO, PbO, Rb_2O
(*iii*) P_4O_{10}, As_4O_6, Li_2O, MnO_2
(*iv*) SO_3, Mn_2O_7, H_2O, Ba_2O

19 Which of the following statements about the series Be, Mg, Ca, Sr and Ba is/are correct?

(*i*) The solubility of the sulphates in water decreases from Be to Ba.
(*ii*) The solubility of the hydroxides in water decreases from Be to Ba.
(*iii*) Calcium carbonate is more stable to heat than barium carbonate.
(*iv*) The polarising power of the Be^{2+} cation is greater than that of the corresponding barium ion.

20 Palladium (Pd) is in the same group of the periodic table as nickel. Which of the following properties of the aqueous Pd^{2+} ion differ from those of the corresponding nickel ion?

(*i*) Pd^{2+}(aq) reacts with H_2S in acid, aqueous solution to give a black precipitate of PdS.
(*ii*) Pd^{2+}(aq) reacts with H_2S in alkaline, aqueous solution to give a black precipitate of PdS.
(*iii*) Pd^{2+}(aq) forms a precipitate with dimethylglyoxime.
(*iv*) Pd^{2+}(aq) forms a precipitate with OH^-(aq) which dissolves in excess of the reagent.

Matching pairs questions

Questions 21–24

Choose from A to E the graph which correctly describes each of the relationships specified in questions 21–24.

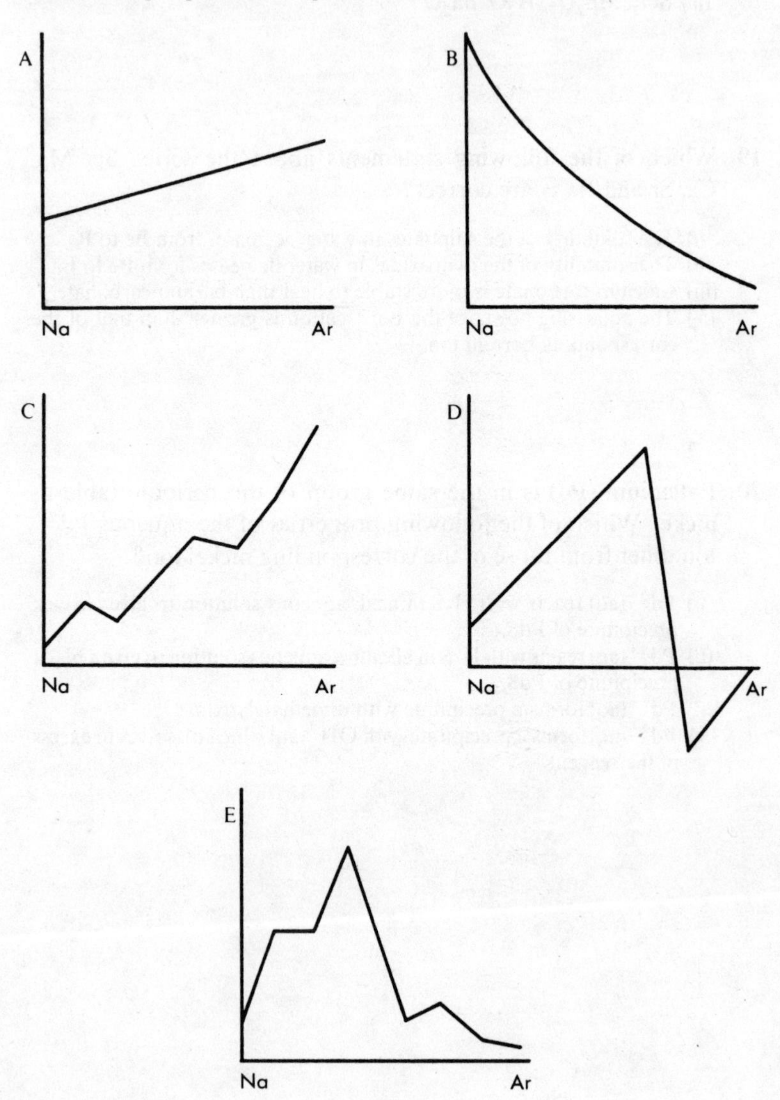

21 The variation in the melting point of the elements from sodium to argon inclusive.

22 The variation in atomic number from sodium to argon inclusive.

23 The variation in principal oxidation state from sodium to argon inclusive.

24 The variation in first ionization energy from sodium to argon inclusive.

Assertion-reason questions

	ASSERTION		REASON
25	Potassium sulphate is isomorphous with potassium selenate	*because*	selenium and sulphur are both in the same group of the periodic table.
26	Elements of atomic number 58–71 (the rare earths) are chemically similar to lanthanum	*because*	these elements are difficult to separate by chemical methods.
27	All nuclides (elements) with an atomic number greater than that of uranium (=92) are radioactive	*because*	they are all synthetic elements which are not found in nature.
28	The chemistry of Tl^+ and Ag^+ shows marked similarities	*because*	thallium and silver atoms each have one electron in the outer shell.
29	The fluorine atom has a higher charge density than the iodine atom	*because*	fluorine is a more powerful oxidizing agent than iodine.
30	^{235}U and ^{238}U may be separated by thermal diffusion of the gaseous isotopes $^{235}UF_6$ and $^{238}UF_6$	*because*	molecules of $^{235}UF_6$ and $^{238}UF_6$ contain different numbers of protons.

9 The Chemistry of the Non-Metals

Multiple choice questions

1 Which one of the following relationships is *incorrect*?

A The electronegativity of phosphorus is greater than that of antimony.
B The melting point of antimony is greater than that of phosphorus.
C Phosphine, PH_3, is more basic than stibine, SbH_3.
D Stibine, SbH_3, is more thermally stable than ammonia, NH_3.
E The first ionization energy of antimony is greater than that of bismuth.

2 Which one of the following statements is correct?

A $Fe(H_2O)_6^{2+}$ is a stronger acid than $Fe(H_2O)_6^{3+}$ in aqueous solution.
B HF is a stronger acid in aqueous solution than HCl.
C BF_3 and SO_3 are examples of "Lewis" bases.
D H_3AsO_3 is a weaker acid in aqueous solution than H_3AsO_4.
E The acid strength of a given compound is independent of the solvent.

3 The process represented by the equation $O(g) + 2e^- = O^{2-}(g)$, is strongly endothermic. Which one of the following helps to explain the existence of a large number of stable metallic, ionic oxides?

A Most metals have large ionization energies.
B Oxygen is too electronegative to form covalent compounds.
C The electron affinities of metals always represent an exothermic process.
D The heat of dissociation of the O_2 molecule is low.
E The lattice energy of the oxides counterbalances the unfavourable electron affinity of oxygen.

4 In which one of the following is there no change in oxidation state?

A the decomposition of chlorine water in sunlight
B the decomposition of an aqueous solution of hydrogen sulphide on standing in air
C the addition of chlorine water to aqueous potassium iodide
D the thermal decomposition of ammonium carbonate
E the thermal decomposition of lead nitrate

5 Which one of the following pairs of elements does *not* form a binary compound?

A arsenic and hydrogen
B fluorine and chlorine
C neon and oxygen
D tin and hydrogen
E xenon and oxygen

6 Which one of the following liberates nitrogen on being heated to 300°C?

A Mg_3N_2
B N_2O_4
C $(NH_4)_2Cr_2O_7$
D NH_4NO_3
E $(NH_4)_2SO_4$

Multiple completion questions

7 Which of the following exist in the solid state as a macromolecule or ionic lattice?

(*i*) SiO_2
(*ii*) PbO_2
(*iii*) SnO_2
(*iv*) CO_2

8 In which of the following can the chemistry of boron be correctly described as differing from that predicted by a knowledge of the chemistry of aluminium?

(*i*) It is a brittle semiconductor.
(*ii*) It forms a number of gaseous hydrides.
(*iii*) It is inert towards oxygen at room temperatures.
(*iv*) It forms an oxide with acidic properties.

9 Hydrogen peroxide in aqueous solution reacts with each of the substances below. In which case(s) does it act as a reducing agent?

(*i*) Ag_2O
(*ii*) $Ba(OH)_2(aq)$
(*iii*) $K_3Fe(CN)_6$ in alkaline solution
(*iv*) $K_4Fe(CN)_6$ in acid solution

10 Which of the following could be used to distinguish $NO_2^-(aq)$ from $NO_3^-(aq)$?

(*i*) acidified potassium permanganate solution
(*ii*) aniline (phenylamine) and dilute hydrochloric acid
(*iii*) warm, dilute hydrochloric acid
(*iv*) iron (II) sulphate and concentrated sulphuric acid

11 Which of the following is/are a correct statement(s) of the properties of the tetrachlorides of carbon and silicon?

(*i*) They act as oxidizing agents.
(*ii*) They form compounds of the type ZCl_6^{2-} where Z is carbon or silicon.
(*iii*) They are readily hydrolysed by warm water.
(*iv*) They are liquids at room temperature.

12 Reactions between hydrocarbons and fluorine are more easily initiated than the corresponding reactions with chlorine. Which of the following helps to explain this fact?

(*i*) A fluorine atom has fewer electrons than a chlorine atom.
(*ii*) The electron affinity of fluorine is greater than that of chlorine.
(*iii*) A fluorine atom is more readily ionized than a chlorine atom.
(*iv*) The F—F bond is weaker than the Cl—Cl bond.

13 In which of the following is the oxidation state of sulphur +6?

(*i*) SO_2Cl_2
(*ii*) SO_3
(*iii*) Na_2SO_4
(*iv*) S_6

14 Cyanogen, $(CN)_2$, is sometimes called a "pseudohalogen" because its chemistry is often analogous to that of the true halogens. Which of the following reactions is consistent with the description "pseudohalogen"?

(*i*) $2Na(s)+(CN)_2(g) = 2NaCN(s)$
(*ii*) $Ag^+(aq)+CN^-(aq) = AgCN(s)$
(*iii*) $2Cu^{2+}(aq)+4CN^-(aq) = 2CuCN(s)+(CN)_2(g)$
(*iv*) $(CN)_2(g)+2O_2(g) = 2CO_2(g)+N_2(g)$

15 Which of the following statements about phosphorus pentachloride is/are correct?

(*i*) The solid forms phosphorus oxychloride on treatment with water.
(*ii*) The solid contains the species PCl_4^+ and PCl_6^-.
(*iii*) The structure of phosphorus pentachloride in the vapour state is trigonal bipyramidal.
(*iv*) The equilibrium $PCl_5(s) \rightleftharpoons PCl_3(l) + Cl_2(g)$ is unaffected by the application of pressure.

16 Which of the following increase(s) in the order He-Ne-Ar-Kr-Xe?

(*i*) The atomic radius of the element.
(*ii*) The first ionization energy of the element.
(*iii*) The density of the element at s.t.p.
(*iv*) The ratio of the specific heats at constant volume and constant pressure.

Matching pairs questions

Questions 17–20

Choose from the list A–E the most appropriate description of the conversions in each of questions 17–20.

A dissociation
B degradation
C disproportionation
D desiccation
E sublimation

17 $CuSO_4 . 5H_2O \rightarrow CuSO_4 . H_2O + 4H_2O$

18 $CH_3CONH_2 \rightarrow CH_3NH_2$

19 $S_8 \rightarrow S_6 + S_2$

20 $NH_4Cl(s) \rightarrow NH_4Cl(g) \rightarrow NH_4Cl(s)$

Questions 21–24

Which of the reagents A to E is most suitable for the volumetric estimation of the ions in questions 21 to 24?

A $HCl(aq)$
B $Na_2S_2O_3(aq)$
C $AgNO_3(aq)$
D $NaOH(aq)$
E $KMnO_4$ (acidified, aqueous solution)

21 $Cu^{2+}(aq)$

22 $Fe^{2+}(aq)$

23 $HCO_3^-(aq)$

24 $NO_2^-(aq)$

Assertion-reason questions

In the production of sulphuric acid from SO_2 and O_2 via the Contact Process:

	ASSERTION.		REASON
25	Theoretical considerations favour a low reaction temperature	*because*	the reaction $2SO_2 + O_2 \rightarrow 2SO_3$ is exothermic.
26	Theoretical considerations favour a low reaction pressure	*because*	increased pressures raise the proportion of SO_3 in the equilibrium mixture.
27	A catalyst of vanadium pentoxide is used	*because*	SO_2 and O_2 will not react in the absence of a catalyst.
28	The catalysis is correctly described as homogeneous catalysis	*because*	the catalyst exists in the solid phase.

29	It is necessary to dry and purify the SO_2 and O_2 before conversion to SO_3	*because*	the sulphuric acid is required to have a high degree of purity.
30	The SO_3 produced is dissolved in concentrated sulphuric acid	*because*	SO_3 is not very soluble in water.

10 The Chemistry of the Metals

Multiple choice questions

1 Which one of the following equations represents a reaction which would be given by lithium but *not* by either sodium *or* potassium? Assume each metal is heated in the gas concerned. (M is lithium, sodium or potassium.)

A $2M + Br_2 = 2MBr$
B $M + O_2 = MO_2$
C $6M + N_2 = 2M_3N$
D $2M + H_2 = 2MH$
E $2M + 2NH_3 = 2MNH_2 + H_2$

2 Which one of the following numerical relationships for stated properties of lithium, sodium and potassium is *not* true?

A the hydration energy $Li^+ > Na^+ > K^+$
B the first ionization energy $Li > Na > K$
C the electronegativities $Li > Na > K$
D the ionic radius $Li^+ < Na^+ < K^+$
E the reduction potentials $M^+(aq)/M$ $Li < Na < K$

3 Which one of the following statements about the elements sodium, potassium, magnesium and calcium is *false*?

A They tarnish on exposure to air.
B The metals are manufactured by electrolysis.
C They are strong reducing agents.
D They form stable, solid bicarbonates.
E The anhydrous sulphates do not decompose below 100°C.

4 Which one of the following compounds in liquid ammonia will produce a reaction analogous to that produced by rubidium oxide, Rb_2O in water?

A Rb_3N
B $RbNH_2$
C RbH
D $RbOH$
E Rb_2O

5 Which one of the following elements does not normally show a stable +1 oxidation state?

A hydrogen
B sodium
C zinc
D mercury
E iodine

6 Which one of the following elements does not normally show a stable +2 oxidation state?

A beryllium
B copper
C silver
D cadmium
E manganese

7 Each element in the alkaline earth "family" (beryllium to radium) has two electrons in its outermost orbital. The element in which these electrons are held most strongly is also

A the weakest reducing agent
B radioactive
C the most chemically reactive
D the weakest electron acceptor
E that with the largest ionic radius

8 Which one of the following correctly explains why calcium and chlorine form the compound $CaCl_2$ rather than the compound $CaCl_3$?

A More energy is required to separate the chlorine molecules in the formation of $CaCl_3$(s).
B More energy is released by the formation of chloride ions from chlorine atoms in the formation of $CaCl_3$(s).
C More energy is required to remove three electrons from the calcium atom than to remove two electrons from the calcium atom.
D More energy is released in the formation of $CaCl_3$(s) from its ions than in the formation of $CaCl_2$(s) from its ions.
E The change $3Ca^{2+}(g) \rightarrow Ca(g) + 2Ca^{3+}(g)$ is exothermic.

9 Which one of the following correctly explains why calcium and chlorine form the compound $CaCl_2$ rather than the compound CaCl?

A More energy is required to separate the chlorine molecules for the formation of $CaCl_2(s)$.
B More energy is released by the formation of chloride ions from chlorine atoms in the formation of $CaCl_2(s)$.
C More energy is required to remove two electrons from the calcium atom than to remove one electron from the calcium atom.
D More energy is released in the formation of $CaCl_2(s)$ from its ions than in the formation of CaCl(s) from its ions.
E The change $2Ca^+(g) \rightarrow Ca(g) + Ca^{2+}(g)$ is endothermic.

10 Which one of the following statements concerning aluminium is *false*?

A It reacts rapidly with 2M sulphuric acid to form aluminium sulphate.
B Its oxide, Al_2O_3, is amphoteric.
C It is an excellent conductor of electricity.
D It dissolves readily in warm, concentrated sodium hydroxide to liberate hydrogen.
E It is the most abundant metal in the earth's crust.

11 Which one of the following species would *not* disproportionate in warm, neutral aqueous solution?

A H_2O_2
B Cu^+
C Ag^+
D Au^+
E MnO_4^{2-}

12 Which one of the following combinations of reagents could *not* give gaseous hydrogen as a major product?

A electrolysis of aqueous sodium chloride
B zinc and concentrated hydrochloric acid
C aluminium and hot, concentrated sulphuric acid
D potassium and methanol
E electrolysis of aqueous potassium acetate

Multiple completion questions

13 Which of the following substances cause(s) the evolution of a gas on addition to warm water?

(*i*) CaC_2
(*ii*) CaH_2
(*iii*) Na_2O_2
(*iv*) Na_2O

14 Which of the following in 0.1M aqueous solution does *not* give a white precipitate when mixed with an equal volume of 0.01M Ba^{2+}(aq) ions?

(*i*) NO_3^-(aq)
(*ii*) CO_3^{2-}(aq)
(*iii*) OH^-(aq)
(*iv*) $C_2O_4^{2-}$(aq)

15 Which of the following statements is correct about the titration of sodamide, $NaNH_2$, with ammonium chloride in liquid ammonia?

(*i*) The titration involves an acid-base reaction.
(*ii*) The electrical conductivity of the solution decreases until the end point is reached.
(*iii*) The amide ion, NH_2^-, acts as a base.
(*iv*) The pH of the solution at the end point is 7.

16 Which of the following, in excess 0.1M aqueous solution, will dissolve silver chloride?

(*i*) KCN
(*ii*) NH_3
(*iii*) $Na_2S_2O_3$
(*iv*) $Na_2Cr_2O_7$

17 Aqueous sodium hydroxide is added to each of the following in 0.1M aqueous solution. Which will give a precipitate soluble in excess alkali?

(*i*) Be^{2+}
(*ii*) Al^{3+}
(*iii*) Sn^{2+}
(*iv*) Mg^{2+}

18 In which of the following respects can the behaviour of ammonium salts be correctly described as similar to that of salts of the alkali metals?

(*i*) behaviour on electrolysis of dilute, aqueous solutions containing a given anion
(*ii*) pH of 0.1M aqueous solutions of a given anion
(*iii*) thermal stability
(*iv*) absence of colour

Matching pairs questions

Questions 19–22

Choose from the list A to E the reagent most suitable to bring about the conversions specified in each of questions 19–22.

A $Cl_2(g)$
B concentrated HNO_3
C $N_2(g)$
D $OCl^-(aq)$
E $HCl(g)$

19 $Sn(s) \rightarrow SnCl_2(s)$

20 $Sn(s) \rightarrow SnO_2(s)$

21 $Mg(s) \rightarrow Mg_3N_2(s)$

22 $Pb^{2+}(aq) \rightarrow PbO_2(s)$

Questions 23–26

For each of questions 23 to 26, choose from the list A to E the element which most accurately fits the description given.

A aluminium
B boron
C calcium
D tin
E bismuth

23 forms a large number of hydrides in which the ratio of electrons to bonds is less than 2:1

24 forms two chlorides which exist in different physical states at room temperature

25 forms a chloride which is deliquescent at room temperature and which vaporises easily to yield molecules containing two metal atoms each

26 combines with nitrogen to form a compound with a diamond like structure

Assertion-reason questions

	ASSERTION		REASON
27	Aluminium chloride dissolves in water to give an acidic solution	*because*	the hydrated aluminium ions, $Al(H_2O)_6^{3+}$ release protons in accordance with the equation $Al[(H_2O)_6]^{3+} = H^+(aq) + [Al\,5H_2O(OH)]^{2+}$
28	Lithium metal does not dissolve like the other alkali metals when added to water	*because*	lithium is the weakest reducing agent among the alkali metals.
29	Anhydrous beryllium chloride is largely covalent in character	*because*	the lattice energy for beryllium chloride is small.
30	Beryllium ions $Be^{2+}(aq)$ are formed when beryllium compounds are dissolved in water	*because*	the hydration of the beryllium ion is a strongly exothermic process.

11 General Inorganic Chemistry

Multiple choice questions

1 0.01 g of a monobasic acid require 5.0 cm^3 of 0.01M NaOH for neutralisation. What is the approximate molecular weight of the acid?

A 20
B 50
C 200
D 250
E 500

2 Which one of the following contains at least one covalent bond in which both electrons are supplied by the same atom?

A SF_6
B $(CH_3CO)_2O$
C $SiCl_4$
D SO_4^{2-}
E $PbCl_4$

3 In which one of the following pairs of compounds is the second named substance more soluble in water than the first?

A CaF_2, $CaCl_2$
B AgF, AgCl
C $KHCO_3$, $NaHCO_3$
D $PbCl_2$, $PbSO_4$
E $Na_3Co(NO_2)_6$, $K_3Co(NO_2)_6$

4 Which one of the following statements about the series N-P-As-Sb-Bi is correct?

A The hydrides increase in stability from NH_3 to BiH_3.
B The electronegativity of the elements increases from N to Bi.
C The basicity of the oxide increases from N_2O_3 to Bi_2O_3.
D The chlorides show increasing covalent character from NCl_3 to $BiCl_3$.
E The relative stability of the higher oxidation state increases from N to Bi.

5 Which one of the following terms correctly describes the slow but spontaneous decomposition of hydrogen peroxide to water and oxygen?

A autocatalysis
B disproportionation
C double decomposition
D oxidation
E reduction

6 Which one of the following in aqueous solution is *not* reduced by aqueous sulphide ions?

A SO_2
B MnO_4^-
C Cu^{2+}
D HNO_3
E Fe^{3+}

7 Which one of the following increases in the order Zn-Cd-Hg?

A the thermal stability of the oxides
B the melting points of the elements
C the solubilities of the metallic sulphides
D the amphoteric nature of the oxides
E the densities of the elements

8 Which one of the following in aqueous solution liberates iodine on treatment with excess aqueous iodide ions?

A OH^-
B Cl^-
C Cu^{2+}
D Zn^{2+}
E NO_3^-

Multiple completion questions

9 Which of the following statements about potassium hexacyanoferrate (III) is/are correct?

(*i*) The co-ordination number of the Fe atom is 9.
(*ii*) The Fe atom has acquired the electronic structure of a krypton atom.
(*iii*) The compound gives a deep blue precipitate with Fe^{3+}(aq) ions.
(*iv*) The compound is an oxidizing agent.

10 In which of the following does hydrogen bonding occur?

(*i*) sodium hydride
(*ii*) acetone (propanone)
(*iii*) the ammonium ion
(*iv*) methanol

11 Which of the following could constitute a conjugate acid-base pair?

(*i*) HCl, NaOH
(*ii*) NH_4^+, NH_2^-
(*iii*) CrO_3, Cr_2O_3
(*iv*) H_3O^+, H_2O

12 Which of the following substances in aqueous solution will have a pH greater than 7?

(*i*) potassium carbonate
(*ii*) ammonium chloride
(*iii*) sodium acetate
(*iv*) aluminium sulphate

13 Orthophosphoric acid has the structure $O{=}P(OH)_3$. Which of the following statements is/are correct?

(*i*) The acid is tribasic.
(*ii*) The acid may be obtained by the hydrolysis of $POCl_3$.
(*iii*) The salt Na_3PO_4, in aqueous solution, has a pH greater than 7.
(*iv*) Orthophosphoric acid is a stronger acid in aqueous solution than sulphuric acid, $O_2{\gg}S(OH)_2$

14 Which of the following elements form(s) at least one hydride which is acidic in aqueous solution?

(*i*) chlorine
(*ii*) phosphorus
(*iii*) silicon
(*iv*) sulphur

15 Which of the following solids normally absorbs moisture from the atmosphere?

(*i*) $CaCl_2 . 6H_2O$
(*ii*) KOH
(*iii*) P_4O_{10}
(*iv*) $Na_2SO_4 . 10H_2O$

Matching pairs questions

Questions 16–19

Choose from the list A–E, the electronic structures corresponding to the elements which correctly answer each of questions 16–19.

A	2	8	8	1	
B	2	8	11	2	
C	2	8	18	7	
D	2	8	18	8	
E	2	8	18	18	8

16 Which element has the lowest first ionization energy?

17 Which element has the highest first ionization energy?

18 Which element is classified as a transition element?

19 Which element most readily forms a stable, unit negative ion?

Questions 20–23

Choose from the list A–E, the element which best fits the descriptions given in questions 20–23.

A aluminium
B lithium
C silicon
D sulphur
E xenon

20 m.p. 181°C; good electrical conductor; density 0.54 g cm^{-3}; forms a chloride in which the element forms a uni-charged cation.

21 m.p. 113°C; poor electrical conductor; density 2.1 g cm^{-3}; forms a stable hydride in which two atoms of hydrogen combine with one atom of the element.

22 m.p. -112°C; poor electrical conductor; density 0.006 g cm^{-3}; forms a fluoride containing four atoms of halogen per molecule.

23 m.p. 659°C; good electrical conductor; density 2.7 g cm^{-3}; forms a chloride which is strongly acidic in water.

Assertion-reason questions

	ASSERTION		REASON
24	Methyl orange is a weak acid	*because*	all indicators are either weak acids or weak bases.
25	At room temperatures and pressures rhombic sulphur is always more stable than monoclinic sulphur	*because*	under these conditions monoclinic sulphur has the lower density.
26	Iodine is more electro-positive than chlorine	*because*	the iodine atom contains more protons than the chlorine atom.
27	In aqueous solution, phosphine is a weaker base than ammonia	*because*	phosphorus is more electronegative than nitrogen.
28	Phenolphthalein is not used to titrate dilute sulphuric acid with aqueous ammonia	*because*	aqueous ammonia is too weak an alkali to effect a colour change with phenolphthalein.
29	Phosphine has a higher boiling point than ammonia	*because*	phosphine has a higher molecular weight than ammonia.
30	Manganese(IV) oxide catalyses the thermal decomposition of potassium chlorate	*because*	the manganese(IV) oxide enables the decomposition to proceed by an alternative reaction mechanism.

12 Hydrocarbons

Multiple choice questions

1 If 20 cm^3 of ethane, C_2H_6, are exploded with 80 cm^3 of oxygen what will be the final volume (in cm^3) of the gas remaining when the products have cooled to their initial room temperature and pressure?

A 10
B 35
C 45
D 50
E 60

2 Each of the names in A–E is a possible name for the hydrocarbon

```
             CH3
              |
      CH3—CH—CH—CH3
                  |
     CH3—CH2—CH—CH3
```

What is its correct I.U.P.A.C. name?

A nonane
B 2-butyl-3-methylbutane
C 2-ethyl-3,4-dimethylpentane
D 2,3,4-trimethylhexane
E 3,4,5-trimethylhexane

3 What is the *total* number of isomeric alkenes of formula C_4H_8?

A 3
B 4
C 5
D 6
E 7

4 An alkene X was treated with ozone and the resulting mixture allowed to react with water and a reducing agent. The only organic product was of molecular formula C_2H_4O. Which one of the following substances could be X?

A $CH_3CH{=}CHCH_3$
B $CH_3CH_2CH{=}CH_2$
C $CH_3CH{=}CHCH_2CH_3$
D $CH_2{=}CH_2$
E $(CH_3)_2C{=}CH_2$

5 In the reaction between hydrogen bromide and $CH_3CH{=}CH_2$, the principal product is $CH_3CHBr.CH_3$ rather than $CH_3CH_2CH_2Br$. Which one of the following correctly interprets this fact?

A The reaction leading to $CH_3CHBr.CH_3$ proceeds via a primary carbonium ion.
B The reaction leading to $CH_3CHBr.CH_3$ proceeds via a secondary carbonium ion.
C $CH_3CHBr.CH_3$ is a more stable compound than $CH_3CH_2CH_2Br$.
D The initial attack on the alkene is via the Br^- ion.
E The reaction leading to the formation of $CH_3CH_2CH_2Br$ is more rapid than that leading to the formation of its isomer.

6 Which one of the following pairs of substances could *not* react together to yield propene as a major product?

A zinc and 1,2-dichloropropane
B propyne and zinc in hydrochloric acid
C 1-chloropropane and alcoholic potassium hydroxide
D 2-chloropropane and alcoholic potassium hydroxide
E propan-1-ol and phosphoric acid

7 Which one of the following substances has the highest boiling point?

A $CH_3(CH_2)_2CH_3$
B $(CH_3)_2CHCH_3$
C $CH_3(CH_2)_3CH_3$
D $(CH_3)_2CHCH_2CH_3$
E $(CH_3)_4C$

8 Which one of the following compounds decolourises both aqueous bromine *and* dilute, aqueous potassium permanganate?

A
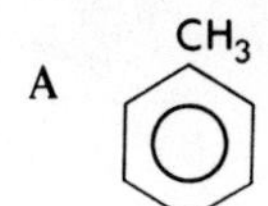

B
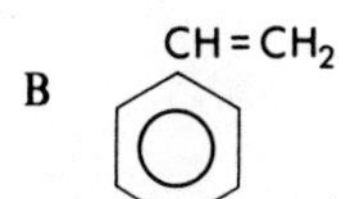

C
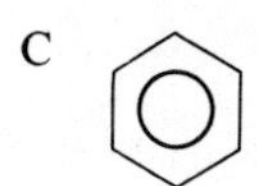

D
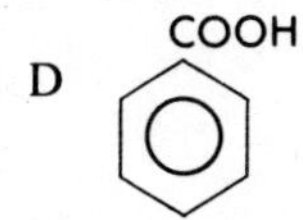

E
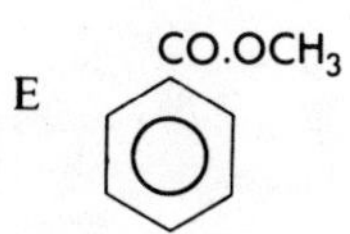

9 Which one of the following compounds has the highest numerical value for its heat of hydrogenation?

A $CH_3CH{=}CHCH{=}CH_2$

B

C
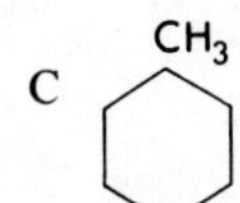

D
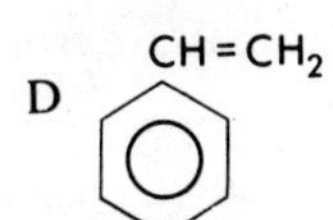

E
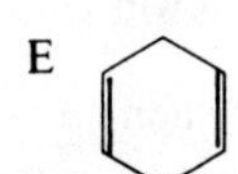

10 Which one of the following is the initial product of the reaction between chlorine and boiling toluene in sunlight?

A
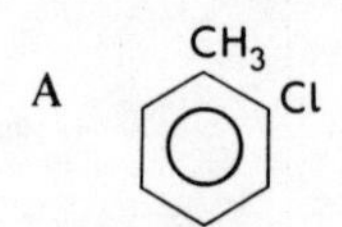

B
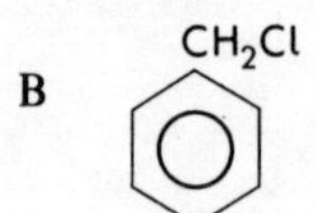

C
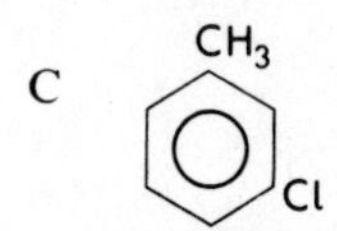

D
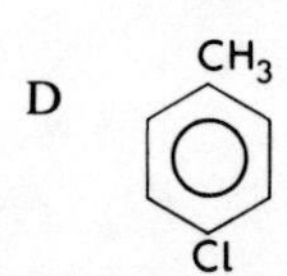

E CH_3, Cl, Cl

Multiple completion questions

11 Which of the following statements about the homologous series of alkanes is/are correct?

(*i*) The general formula is C_nH_{2n+2}.
(*ii*) The difference between successive members corresponds to 14 molecular weight units.
(*iii*) The higher members of the series are solids.
(*iv*) The series cannot contain any optical isomers.

12 Which of the following statements about benzene is/are true?

(*i*) The molecule is planar.
(*ii*) The C—C bond is greater in length than the C—C bond in ethane.
(*iii*) Benzene undergoes substitution reactions when it reacts with chlorine in ultra-violet light.
(*iv*) Benzene may react with bromine additively to form a non-aromatic compound.

13 In which of the following pairs of compounds are the substances isomeric with each other?

(*i*) benzene and 2,4,6-hexatriene
(*ii*) cyclopropane and propene
(*iii*) styrene (phenylethene) and ethyl benzene
(*iv*) hexene and cyclohexane

14 Which of the following compounds would (*a*) form ethanol with warm, dilute aqueous sodium hydroxide *and* (*b*) decolourise bromine dissolved in acetic acid?

(*i*) $CH_3CH_2CO.CH_2OCH{=}CH_2$
(*ii*) $CH_3CH_2CH{=}CH.CH{=}CHOCH_2CH_2OH$
(*iii*) $CH_3CH_2COOCH_2CH_2CH{=}CHCH_3$
(*iv*) $CH_3CH{=}CHCH_2CH_2COOCH_2CH_3$

15 Which of the following products could be obtained by the electrolysis of aqueous potassium *n*-butyrate using platinum electrodes?

(*i*) carbon dioxide
(*ii*) propane
(*iii*) oxygen
(*iv*) *n*-hexane

16 Which of the following compounds will decolourise aqueous, alkaline potassium permanganate?

(*i*) 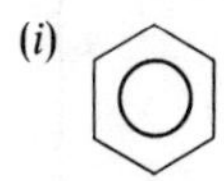

(*ii*) 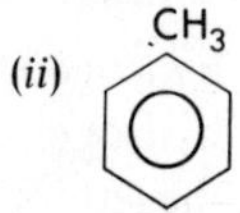

(*iii*) $CH_3CH_2CH_3$
(*iv*) $CH_2{:}CHCH_3$

Matching pairs questions

Questions 17–20

Choose from the list A–E the term which correctly describes the relationship between members of the pairs of compounds in questions 17–20.

A conjugate acid-base pair
B enantiomers
C geometric isomers
D functional isomers
E positional isomers

17 $CCl_2{=}CH_2$ and $CHCl{=}CHCl$

18 CH_3CH_2COOH and CH_3COOCH_3

19 CH_3CH_2CHO and CH_3COCH_3

20 $HC{\equiv}CH$ and $HC{\equiv}C^-$

Questions 21–23

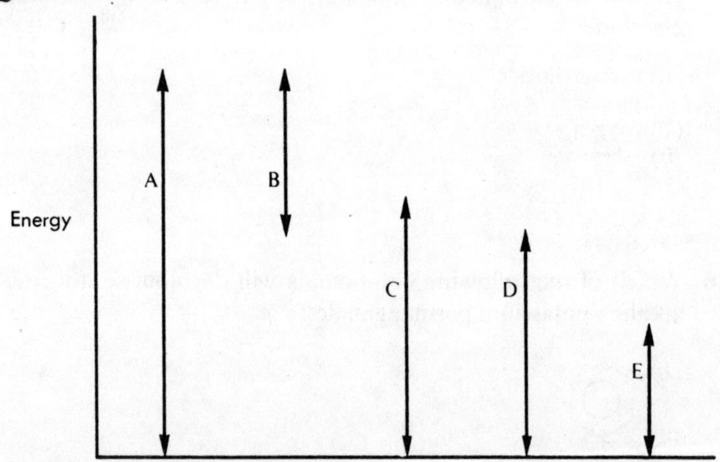

The above diagram shows the relative molar heats of hydrogenation of a number of hydrocarbons, the product of the hydrogenation in each case being cyclohexane, C_6H_{12}. Choose from the letters A–E the energy change corresponding to the hydrogenation of each of the compounds in questions 21 to 23.

21 , C_6H_6

22 , C_6H_8

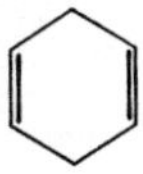

23 , C_6H_{10}

Assertion-reason questions

	ASSERTION		REASON
24	Petroleum may not be cracked in the absence of a catalyst	*because*	the reactions which take place in cracking involve homolytic fission of covalent bonds.
25	The ratio of the C≡C to C=C bond energies is 3:2	*because*	the ratio of the number of electrons involved in the bonds is also 3:2
26	The principal product of the reaction between $(CH_3)_2C{=}CH_2$ and HCl is $(CH_3)_3CCl$ rather than $(CH_3)_2CHCH_2Cl$	*because*	tertiary carbonium ions are more stable than primary carbonium ions.
27	Vinyl chloride, $CH_2{=}CHCl$ is less reactive than ethyl chloride towards nucleophilic substitution	*because*	free rotation about the C=C bond is not possible.
28	Sulphonation of toluene is more rapid than sulphonation of benzene under comparable conditions	*because*	the methyl group in toluene activates the benzene ring.

29	Acetylene polymerises at red heat to form benzene	*because*	acetylene is an endothermic compound.
30	Naphthalene is a covalent compound	*because*	it melts below 100°C at 1 atm. pressure

13 Compounds of Carbon, Hydrogen and Oxygen

Multiple choice questions

1 Which one of the following substances has the lowest boiling point?

A $CH_3CH_2CH_2CH_2OH$
B $(CH_3)_2CHCH_2OH$
C $(CH_3)_3COH$
D $CH_3CH_2CHOH.CH_3$
E $CH_3CH{=}CHCH_2OH$

2 The formula of glucose may be written as

$$CH_2OH.CHOH.CHOH.CHOH.CHOH.CHO$$

How many asymmetrically bonded carbon atoms are there per molecule?

A 2
B 3
C 4
D 5
E 6

3 Which one of the following correctly represents the order of carbon-oxygen bond lengths in (*i*) carbon dioxide, (*ii*) dimethyl ether and (*iii*) the acetate ion?

A (*i*) > (*ii*) > (*iii*)
B (*i*) > (*iii*) > (*ii*)
C (*ii*) > (*i*) > (*iii*)
D (*ii*) > (*iii*) > (*i*)
E (*iii*) = (*ii*) > (*i*)

4 Which one of the following carboxylic acids has the lowest first ionization constant?

A CH_3COOH
B CH_3CH_2COOH
C $CH_2Cl.COOH$
D $(COOH)_2$
E $HOOCCH_2COOH$

5 Which one of the following produces the most strongly acidic solution in water?

A $ClCH_2COONa$
B CH_3COCl
C $C_6H_5CH_2Cl$
D C_6H_5OH
E $ClCH_2COOCH_2CH_3$

6 The alkene hex-3-ene is treated with ozone and the product hydrolysed. Which of the following would be the cleavage products of this reaction?

A acetaldehyde (ethanal) and acetone (propanone)
B propionaldehyde (propanal) and propanol
C acetone (propanone) only
D propionaldehyde (propanal) only
E acetaldehyde (ethanal) only

7 Which one of the following would you use to distinguish a simple aldehyde from a simple ketone?

A hydroxylamine
B 2,4-dinitrophenylhydrazine
C aqueous potassium permanganate
D hydrogen chloride
E sodium bisulphite

8 Which one of the following could *not* be prepared directly from ethanal, CH_3CHO, by a one stage synthesis?

A CH_3CH_2COOH
B $CH_3CHOH.CH_2CHO$
C $CH_3CH{=}N.NHC_6H_5$
D CHI_3
E $CH_3CH{=}NOH$

Multiple completion questions

9 Which of the following has (have) neither optical nor geometric isomers?

(*i*) COOH
|
COOH

(*ii*) $CH_3CH(OH).CH(OH).CH_3$

(*iii*) $CHCl{=}CHCl$

(*iv*) $C_6H_5CO.C_6H_4Cl$

10 Which of the following molecules is/are planar?

(*i*) CH_2OH

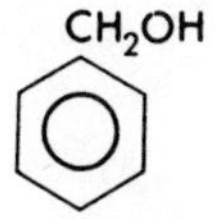

(*ii*)

(*iii*) CH_3

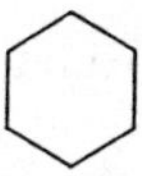

(*iv*) $CH{=}CH_2$

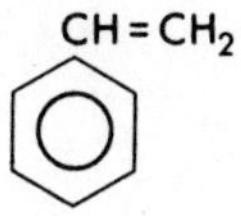

11 Which of the following substances would you expect to be present if ethanol is allowed to react with concentrated sulphuric acid?

(*i*) $(C_2H_5)_2O$

(*ii*) C_2H_4

(*iii*) $C_2H_5HSO_4$

(*iv*) $(C_2H_5)_2C{=}O$

12 Which of the following in aqueous solution would you expect to be a stronger acid than phenol (benzenol)?

(*i*)

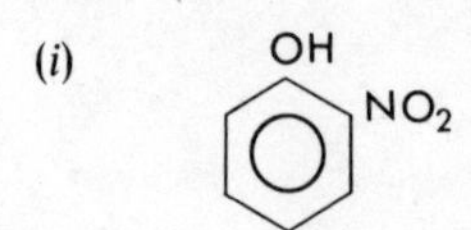

(*ii*)

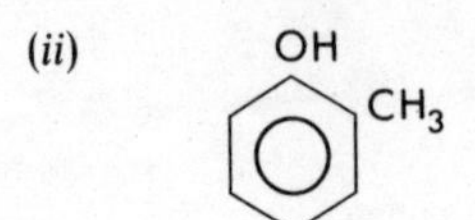

(*iii*)

OH
NO_2 NO_2
NO_2

(*iv*)

OH
CH_3

13 Which of the following compounds, after oxidation with aqueous potassium permanganate, will form an acid which on heating yields an anhydride?

(*i*)

CH_3
CH_3

(*ii*)

COOH
CHO

(*iii*)

COOH
OH

(*iv*)

CH_3
CH_2CH_3

14 Which of the following compounds yields benzoic acid (benzene-carboxylic acid) on prolonged boiling with excess, alkaline potassium permanganate?

(i)

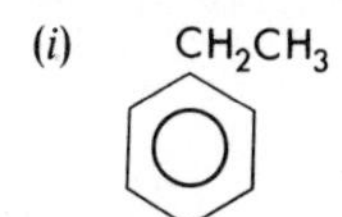

(ii)

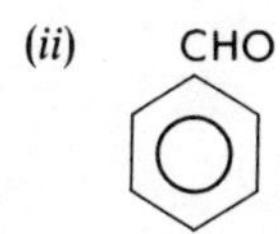

(iii)

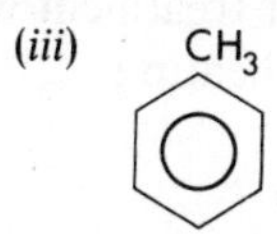

(iv)

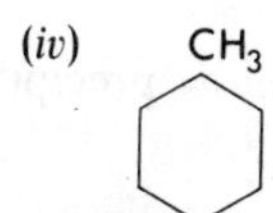

Matching pairs questions

Questions 15–18

Choose from the list A–E the compound which correctly accounts for the properties outlined in each of questions 15–18.

A CH_3COCH_3
B $CH_2Br.CH_2OH$
C CH_3OCH_3
D $CH_3CH_2COOCH_3$
E CH_3CH_2CHO

15 obtained by heating isopropanol (propan-2-ol) with excess acidified potassium dichromate solution

16 forms ethylene glycol (ethane-1,2-diol) on refluxing with aqueous potassium hydroxide

17 reacts with sodium metal to liberate hydrogen

18 reduces Fehling's or Benedict's solutions and ammoniacal silver nitrate solution

Questions 19–22

Choose from the list A–E the compound which corresponds to each of the descriptions given in questions 19–22.

A HCHO
B CH_3CHO
C C_6H_5CHO
D CH_3COOH
E CH_3CH_2OH

19 does not react with phosphorus pentachloride

20 forms the compound $CH_3CH{=}CH.CHO$ after treatment with dilute, aqueous sodium carbonate followed by heating

21 does not react with aqueous sodium hydroxide

22 is not readily oxidized and does not give a yellow precipitate with iodine in aqueous, alkaline solution

Assertion-reason questions

	ASSERTION		REASON
23	Dimethylether (methoxymethane) has a lower boiling point than ethanol	*because*	it does not contain a highly polar OH group.
24	Propionic acid (propanoic acid) is more volatile than methyl acetate (methyl ethanoate) which has the same molecular weight	*because*	the intermolecular forces of repulsion are greater in the acid than in the ester.
25	$CH_3CH_2CH_2COOH$ and $CH_3COOCH_2CH_3$ have the same molecular weight	*because*	they are geometric isomers.

26	Hydrogen cyanide reacts more rapidly with acetone (propanone) in the presence of a little aqueous alkali	*because*	the initial step of the reaction involves nucleophilic attack by cyanide ions.
27	In a reaction involving nucleophilic attack on a molecule, the reaction will be more rapid in the presence of $OH^-(aq)$ than $H^+(aq)$	*because*	$OH^-(aq)$ is a stronger nucleophile than $H^+(aq)$.
28	Potassium bromide does not react with ethanol in the absence of sulphuric acid	*because*	the ethanol molecules must first be protonated before bromide ions can attack them.
29	Formaldehyde (methanal) reacts more rapidly with cyanide ions than does acetone (propanone)	*because*	formaldehyde is a gas at room temperature whereas acetone is a liquid.
30	Glucose will reduce Fehling's or Benedict's solution	*because*	glucose is a carbohydrate.

14 The Organic Chemistry of Nitrogen and the Halogens

Multiple choice questions

1 Which one of the following classes of organic compounds cannot be prepared from alkyl halides by a single synthesis?

A alcohols
B alkanes
C aldehydes
D alkylbenzenes
E ethers

2 Which one of the following compounds does *not* yield ammonia on boiling with excess, aqueous sodium hydroxide?

A CH_3CONH_2
B CH_3CH_2CN
C C_6H_5CN
D $CH_3CH_2NH_2$
E CH_3COONH_4

3 Which one of the following statements concerning alanine (2-aminopropanoic acid) $CH_3CHNH_2.COOH$ is *false*?

A Optical isomers of this formula exist.
B A sample extracted from naturally occurring protein is optically active.
C The acid dissociation constant is higher than that of the unsubstituted carboxylic acid CH_3CH_2COOH.
D The basic dissociation constant is lower than that of NH_3.
E Internal rearrangement will tend to give a polar structure $^{+}NH_3CH(CH_3)COO^{-}$.

4 What is the maximum number of species which could theoretically be formed by the progressive substitution of chlorine atoms into an ethane molecule?

A 6
B 7
C 8
D 9
E 10

5 If CH_3CH_2Y reacts with CH_3I to give $(CH_3CH_2YCH_3)^+I^-$, what is the identity of Y?

A OH
B NH_2
C $NHCH_3$
D $N(CH_3)_2$
E CN

6 How many *additional* isomers of the compound

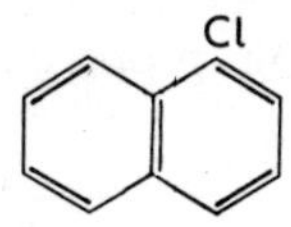

may be obtained by moving the chlorine atom to other positions in the naphthalene rings?

A 1
B 2
C 4
D 5
E 8

Multiple completion questions

7 Which of the following alcohols will give iodoform (triodomethane) on treatment with iodine and aqueous sodium hydroxide?

(*i*) CH_3OH
(*ii*) $CH_3CH_2CH_2OH$
(*iii*) $(CH_3)_3COH$
(*iv*) $CH_3CHOHCH_3$

8 Which of the following will *not* produce a precipitate on prolonged warming with alcoholic silver nitrate?

(*i*) CH_3CH_2Br
(*ii*) $CH_2{=}CHBr$
(*iii*) $C_6H_5CH_2Br$
(*iv*) C_6H_5Br

9 Which of the following statements is/are correct?

(*i*) Alkyl iodides hydrolyse more readily than alkyl chlorides.
(*ii*) Cu(II) can be reduced to Cu(I) by I^-(aq) ions.
(*iii*) Iodine in alkaline solution gives a yellow precipitate on warming with ethanol.
(*iv*) Iodine dissolved in aqueous potassium iodide oxidizes Hg(II) to Hg(IV).

10 Which of the following properties would you expect of nicotine?

N N CH_3

(*i*) It would exhibit basic properties.
(*ii*) It would react with boiling, acidified potassium permanganate solution.
(*iii*) It would form a quaternary ammonium salt.
(*iv*) It will form a diazonium compound with cold dilute hydrochloric acid and sodium nitrite.

11 Which of the following substances give rise to enantiomers (optical isomers)?

(*i*) 1,2-dichlororoethene, $CHCl{=}CHCl$
(*ii*) 2,3-dichlorobutane, $CH_3CHCl.CHClCH_3$
(*iii*) 1-chloro, 2-iodoethene, $CHI{=}CHCl$
(*iv*) lactic acid (2-hydroxypropanoic acid), $CH_3CHOH.COOH$

12 For which of the following pairs is the first named substance a reagent which could be used to prepare the second named substance from a benzenediazonium salt?

(*i*) $NaBF_4$, C_6H_5F
(*ii*) KCl, C_6H_5Cl
(*iii*) KBr, C_6H_5Br
(*iv*) KI, C_6H_5I

Matching pairs questions

Questions 13–16

Choose from the list A–E the reagent or reagents which would be used to effect the conversions indicated in each of questions 13–16.

A $HCl(aq)$
B $Cl_2(g)$ and ultra-violet light
C $COCl_2(g)$ and anhydrous aluminium chloride
D $CHCl_3(l)$ and anhydrous aluminium chloride
E $PCl_3(l)$

13 OH → Cl

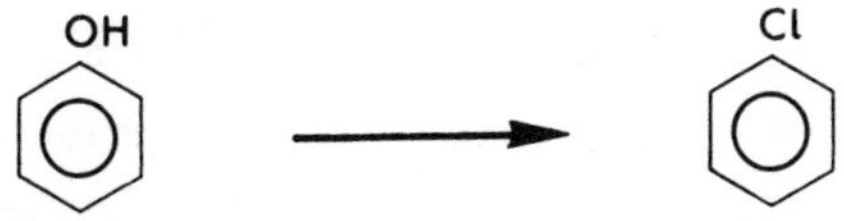

14 → Cl Cl Cl Cl Cl Cl

15 → C=O

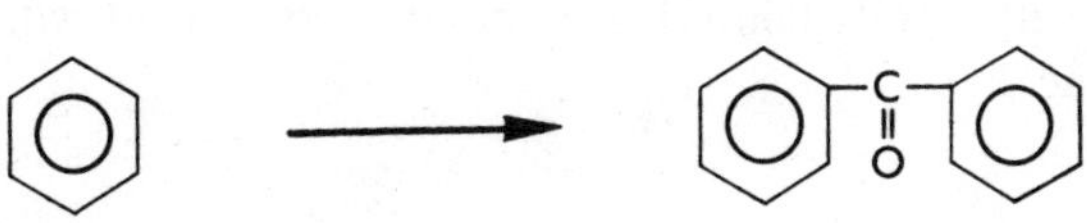

16 → CH 3

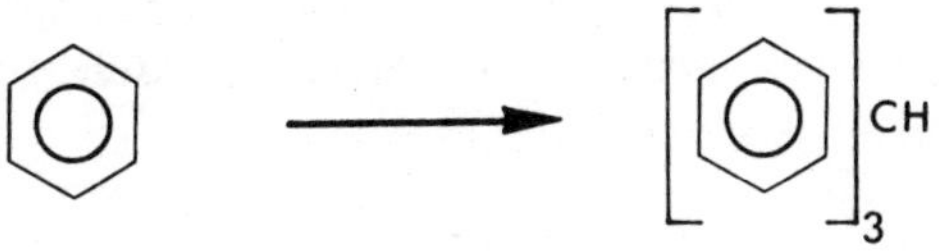

Questions 17–20

Choose from the list A–E the compounds that undergo the reactions described in each of questions 17–20.

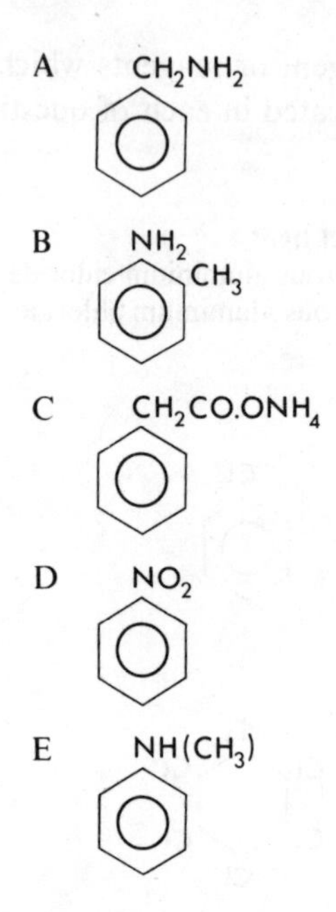

17 forms a yellow, oily neutral compound on treatment with cold nitrous acid

18 forms a diazonium salt on treatment with cold nitrous acid

19 forms ammonia when heated with excess aqueous sodium hydroxide

20 forms a mixture of alcohols and alkanes and liberates nitrogen on treatment with cold nitrous acid

Questions 21–24

Choose from the list A–E the description which most accurately indicates the mechanism of the change given in questions 21–24.

A electrophilic addition
B electrophilic substitution
C nucleophilic addition
D nucleophilic substitution
E free radical substitution

21 $C_6H_5OH \xrightarrow{Br_2(aq)} C_6H_2Br_3OH$ (phenol → 2,4,6-tribromophenol: OH at top; Br, Br, Br at the 2-, 4- and 6-positions)

22 $CH_3CH{=}CH_2 \xrightarrow{HCl} CH_3CHCl.CH_3$

23 $CH_3CH_2CH_2Cl \xrightarrow[(aq/alc)]{CN^-} CH_3CH_2CH_2\ CN$

24 $CH_3CH_2CH_3 \xrightarrow[u.v.\ light]{Br_2(g)} CH_3CHBr.CH_3$

Assertion-reason questions

	ASSERTION		REASON
25	Steam distillation allows aniline (phenylamine) to be distilled below its normal boiling point	*because*	the water increases the partial vapour pressure of the aniline at 100°C.
26	Amines form salts more readily than amides	*because*	there is no lone pair of electrons on the nitrogen atom in amides.
27	Nylon is a polyamide	*because*	it is made from a nitrogen containing acid.

28	Chloroform (trichloromethane) yields sodium formate (methanoate) on heating with excess aqueous sodium hydroxide	*because*	chloroform behaves as the acid chloride of formic (methanoic) acid.
29	Methyl iodide (iodomethane) reacts with methanol to form diethyl ether (ethoxyethane)	*because*	the oxygen atom in methanol is able to displace the iodine atom from the alkyl halide.
30	Aniline is a stronger base than methylamine	*because*	the benzene ring is electron donating.

15 General Organic Chemistry

Multiple choice questions

1 One mole of each of the following is completely oxidized. Which one would yield 4 moles of water?

A ethanol
B methanol
C propane
D butane
E buta-1,3-diene

2 Which one of the following is the correct formula for 1,1-diethylethene?

A $CH_3CH_2\underset{CH_3}{\underset{|}{C}}HCH_2CH_3$

B $CH_3\underset{CH_2CH_3}{\underset{|}{C}}H{-}CH{=}CH_2$

C $CH_3CH_2\underset{CH_2CH_3}{\underset{|}{C}}{=}CH_2$

D $H\underset{CH_2CH_3}{\underset{|}{C}}{=}CHCH_2CH_3$

E $CH_3CH_2CH{=}CHCH_2CH_3$

3 Which one of the following would you expect to have the lowest boiling point under given conditions?

A $CH_3CH_2CH(CH_3)_2$
B $CH_3CH_2CH_2CH_2COOH$
C $CH_3CH_2CH_2CH_2CH_3$
D $CH_3CH_2CH_2CH_2CH_2Br$
E $CH_3CH_2CH_2CH_2CH_2NH_2$

4 One mole of triphenylcarbinol $[(C_6H_5)_3COH]$ lowers the freezing point of 1 000 g of sulphuric acid twice as much as one mole of methanol. Which one of the following correctly explains this fact?

A Methanol, unlike triphenylcarbinol, dissociates in sulphuric acid.
B Triphenylcarbinol is dissociated into $(C_6H_5)_3C^+$ and OH^- ions whereas methanol is unchanged.
C Triphenylcarbinol dissociates to form $(C_6H_5)_3C^+, H_3O^+$ and $2HSO_4^-$ ions whereas methanol forms only CH_3^+ and OH^-.
D Methanol combines with sulphuric acid to form CH_3HSO_4 and H_2O whereas triphenylcarbinol is unchanged.
E Triphenylcarbinol dissociates to form $(C_6H_5)_3C^+, H_3O^+$ and $2HSO_4^-$ ions whereas methanol forms only $CH_3OH_2^+$ and HSO_4^-.

5 Compounds $CH_3CH_2CH_2Y$ and $CH_3CH_2C{\lesssim}^{Y}_{O}$ are both slightly soluble in water and neither is easily oxidized. Which one of the following could Y represent?

A Cl
B NH_2
C OH
D OCH_3
E H

6 Which one of the following reacts readily with bromine water to form a substitution product?

A

B OH (benzene ring)

C NO_2 (benzene ring) NO_2

D

E CH_3 (benzene ring)

7 Which one of the following compounds (*i*) reacts with sodium hydroxide but not with sodium bicarbonate *and* (*ii*) forms a precipitate with aqueous bromine?

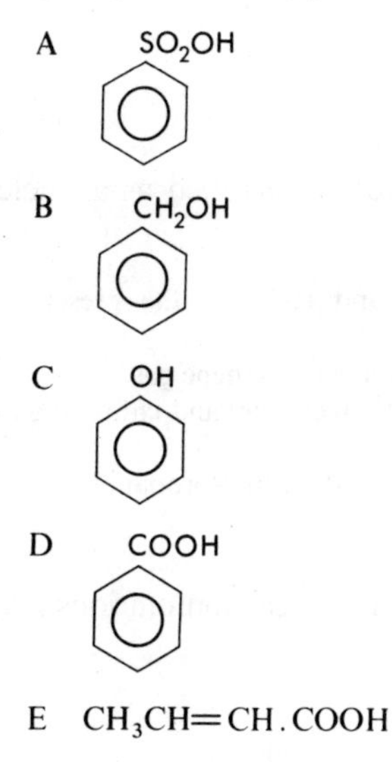

E $CH_3CH{=}CH.COOH$

8 Which one of the following correctly defines the industrial term "heavy chemical"?

A one made in large quantities
B one of high molecular weight
C one which is expensive to manufacture
D one which is in short supply
E one of high density

Multiple completion questions

9 A compound contain 85.6% carbon and 14.4% hydrogen by weight. Which of the following statements about the compound *must* be correct?

(*i*) It has an empirical formula CH_2.
(*ii*) It is a gas at room temperature.
(*iii*) It is an unsaturated compound.
(*iv*) It is an aliphatic compound.

10 Which of the following molecules is/are planar?

(*i*) formaldehyde, methanal, HCHO
(*ii*) ethene, CH_2CH_2
(*iii*) ethyne, HC⋮CH
(*iv*) propene, CH_3CHCH_2

11 Which of the following is/are correctly classified as a nucleophilic substitution?

(*i*) The reaction between CH_3CHO and HCN in the presence of OH^-(aq).
(*ii*) The reaction between aqueous bromine and ethene gas.
(*iii*) The reaction between xylene (dimethylbenzene) and chlorine gas in the presence of sunlight.
(*iv*) The reaction of acetyl (ethanoyl) chloride with *n*-propanol.

12 Which of the following statements about carbonium ions is/are correct?

(*i*) They are free radicals.
(*ii*) They contain a carbon atom with only six electrons.
(*iii*) A tertiary carbonium ion is more stable than a primary carbonium ion.
(*iv*) There is no evidence for their existence because of their very short life.

13 Which of the following exhibit(s) geometrical isomerism?

(*i*) but-2-ene
(*ii*) but-2-yne
(*iii*) 1,2-dibromoethene
(*iv*) phthalic acid, $C_6H_4(COOH)_2$

14 Which of the following statements is/are correct?

(*i*) Aryl halides do not undergo nucleophilic substitution.
(*ii*) The presence of an unshared electron pair is characteristic of a nucleophile.
(*iii*) In mono-nitrobenzene, ortho- and para-substitution is more rapid than meta-substitution under given conditions.
(*iv*) *p*-nitrobenzoic acid is a stronger acid in aqueous solution than benzoic acid itself.

15 In which of the following base-pairs is the first named the stronger base in aqueous solution?

(*i*) methylamine and ammonia
(*ii*) methylamine and dimethylamine
(*iii*) methylamine and aniline
(*iv*) diphenylamine and aniline

16 In which of the following reactions does the first substance specified in the equation act as an acid?

(*i*) $C_6H_5COCl + H_2O = C_6H_5COOH + HCl$
(*ii*) $CH_3COCH_3 + HCN = (CH_3)_2C(OH)CN$
(*iii*) $C_2H_6 + Cl_2 = C_2H_5Cl + HCl$
(*iv*) $HBr + C_6H_5CH{=}CH_2 = C_6H_5CHBr.CH_3$

17 Which of the following pairs of substances are isomeric monosaccharides?

(*i*) sucrose and maltose
(*ii*) maltose and glucose
(*iii*) starch and sucrose
(*iv*) glucose and fructose

18 Which of the following statements about starch and cellulose is/are correct?

(*i*) Both can be hydrolysed to glucose.
(*ii*) Both can be digested by humans.
(*iii*) Both give a blue colour with aqueous iodine.
(*iv*) Both are polysaccharides.

Matching pairs questions

Questions 19–22

Choose from the list A–E the reagent most suitable for carrying out the conversions in questions 19–22.

A $Br^-(aq)$
B conc. $HBr(aq)$
C $Br_2(aq)$
D $Br_2(l)$ in sunlight
E $Br_2(l)$ in the presence of iodine

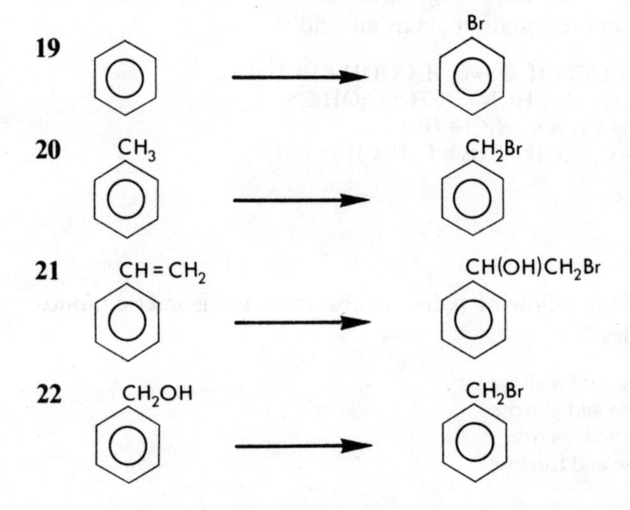

Questions 23–26

Choose from the list A–E the compound which best fits the properties outlined in each of questions 23–26.

A chloroform (trichloromethane)
B acetanilide (*N*-phenylethanamide)
C ammonium acetate (ethanoate)
D acetaldehyde (ethanal)
E ethylene glycol (ethane-1,2-diol)

23 may be easily oxidized to a dicarboxylic acid

24 forms isopropyl alcohol (propan-2-ol) when reacted with methylmagnesium iodide followed by hydrolysis

25 forms an amide on heating

26 forms sodium formate on heating with excess aqueous sodium hydroxide

Assertion-reason questions

	ASSERTION		REASON
27	A racemic mixture has no effect upon plane polarized light, *i.e.* is optically inactive	*because*	racemic mixtures contain equal numbers of molecules of enantiomers.
28	Styrene (phenylethene) may be polymerized by the addition of hydrogen peroxide	*because*	styrene is an unsaturated hydrocarbon.
29	There are two isomeric forms of propyl iodide (iodopropane)	*because*	propyl iodide contains an asymmetrically bonded carbon atom.
30	A soap is produced by the hydrolysis of an ester	*because*	a soap is a salt of a carboxylic acid.

16 Experimental Procedures

Multiple choice questions

1 Which one of the following procedures could be appropriately used to obtain a sample of barium chloride from a solid mixture of barium chloride and barium sulphate?

A heating in an apparatus designed to collect a sublimate
B addition of aqueous ammonia followed by filtration
C addition of excess, aqueous silver nitrate to a solution of the mixture
D addition of warm water, filtration and crystallisation of the filtrate
E addition of warm water, followed by excess, aqueous ammonium carbonate

2 Why is it unsatisfactory experimental procedure to try to condense a liquid boiling at 190°C (1 atm. pressure) with a water-cooled (Liebig) condenser?

A the vapour may condense before it reaches the condenser
B rapid condensation may block the condenser
C the liquid might react with the water
D the water in the condenser might boil
E the condenser might break

3 The sulphonation of benzene requires heat and prolonged contact between benzene and concentrated sulphuric acid over a period of approximately 24 hours. Which one of the following pieces of apparatus would be most appropriate to bring about the reaction between benzene and concentrated sulphuric acid?

A a round bottomed flask fitted with a rubber stopper
B a flask fitted with an air condenser in the vertical position
C a flask fitted with a water-cooled condenser in the horizontal position
D a flask fitted with a water-cooled condenser in the vertical position
E a flask fitted with a tap funnel containing concentrated sulphuric acid

4 A group of students determined the density of a liquid by running it from a Grade A 50 cm^3 burette into a weighing bottle. An analytical balance was used for all weighings. Which one of the following recorded results (in $g\,cm^{-3}$) represents the maximum degree of accuracy possible with this procedure?

A 13
B 13.5
C 13.54
D 13.543
E 13.5427

5 Which one of the following compounds has the highest melting point?

A CH_3COOH
B $CHCl_3$
C C_6H_5COCl
D CH_3CONH_2
E $CH_3COOC_2H_5$

6 When solutions of the substances in each of the following pairs are mixed, in which case will no visible change occur?

A $BaCl_2 + K_2SO_4$
B $FeSO_4 + HNO_3$
C $KOH + MgCl_2$
D $AgNO_3 + FeCl_3$
E $AgNO_3 + NaF$

7 Equal volumes of *equimolar* solutions of each of the following pairs of substances are mixed. In which case is the electrical conductivity of the resulting mixture greatest?

A NaCl and $AgNO_3$
B HCN and $AgNO_3$
C HCl and $AgNO_3$
D $Ba(NO_3)_2$ and H_2SO_4
E $(CH_3COO)_2Pb + H_2SO_4$

8 Iodoethane is sometimes prepared by the action of iodine on ethanol in the presence of red phosphorus. Which one of the following procedures is likely to give the best yield of iodoethane?

A dissolving the iodine in aqueous potassium iodide
B using a large excess of ethanol
C heating the reaction mixture under reflux conditions
D addition of a little aqueous hydrogen iodide
E allowing the reaction mixture to stand for several hours

9 Which one of the following would be best used to separate a small sample of two immiscible liquids with boiling points of 84° and 135°C respectively?

A a retort
B a reflux condenser and a distilling flask
C an air condenser and a distilling flask
D a fractionating column and a distilling flask
E a separating funnel

10 In which of the following measurements is the *percentage* uncertainty the greatest?

A volume of titrant = $(28.35 \pm 0.05)\,cm^3 - (2.35 \pm 0.05)\,cm^3$
B mass of a container = $15.35\,g \pm 0.01\,g$
C mass of precipitate = $(2.1315 \pm 0.0001)\,g - (1.0005 \pm 0.0001)\,g$
D mass of a vehicle = $1260\,Kg \pm 10\,Kg$
E temperature rise = $(27.1 \pm 0.1)°C - (21.2 \pm 0.1)°C$

Multiple completion questions

11 An ethereal solution of a solid S is allowed to form a separate layer on top of a water layer in which S is insoluble. Which of the following will occur over a period of several days?

(*i*) Substance S will remain entirely in ethereal solution.
(*ii*) The two solvents will gradually intermix.
(*iii*) Solute S will partition itself between the two solvents.
(*iv*) Crystals of S will appear at the interface between the two solvents.

12 A student standardising a solution of potassium permanganate using a standard solution of iron(II) obtained a value for the molarity of the permanganate which was subsequently shown to be low. Which of the following could reasonably account for the erroneous result?

(*i*) using distilled water for the final burette rinsing
(*ii*) using distilled water for the final pipette rinsing
(*iii*) carrying out the titration at too low a temperature
(*iv*) using only the first titration value for the estimation

13 Which of the following pieces of apparatus is *essential* for the determination of the standard electrode potential for the Cu^{2+}(aq)/Cu half cell?

(*i*) a Zn^{2+}(aq)/Zn half cell
(*ii*) a voltmeter of some sort or a potentiometer
(*iii*) a burette
(*iv*) an analytical balance

14 In gravimetric analysis rapid precipitation is usually avoided. Which of the following is/are possible reasons for this?

(*i*) Rapid precipitation may yield a precipitate which is hard to collect.
(*ii*) Rapid precipitation may yield a precipitate which is hard to filter.
(*iii*) Rapid precipitation may lead to small crystals of precipitate.
(*iv*) Rapid precipitation may be incomplete.

15 Which of the following may be determined by a single experiment?

(*i*) the heat of combustion of *n*-butanol
(*ii*) the lattice energy of magnesium oxide
(*iii*) the first ionization energy of xenon
(*iv*) the heat of formation of methane

16 Which of the following is/are essential characteristics of a solvent to be used for recrystallisation purposes?

(*i*) immiscibility with water
(*ii*) a pure substance, not a mixture
(*iii*) a low boiling point
(*iv*) able to dissolve more solute when hot than when cold

17 Which of the following procedures could appropriately be used to obtain a sample of beryllium chloride from a mixture containing beryllium chloride and sodium chloride?

(*i*) fractional crystallisation from water
(*ii*) steam distillation
(*iii*) addition of hot water followed by filtration
(*iv*) ether extraction

18 Which of the following methods may be employed to prepare a sample of lead(II) chloride?

(*i*) passing dry chlorine over heated lead
(*ii*) adding warm 2M hydrochloric acid to lead
(*iii*) adding silver chloride to aqueous lead nitrate
(*iv*) adding 2M hydrochloric acid to aqueous lead nitrate

Matching pairs questions

Questions 19–22

Choose from the list A–E the reagent most suitable for precipitating the first named ion only when added to a mixture containing the ion pairs in each of questions 19–22.

A excess NH_3(aq)
B excess HCl(aq)
C excess HNO_3(aq) followed by H_2S(g)
D excess of a mixture of NH_4Cl(aq) and $(NH_4)_2CO_3$(aq)
E excess $(NH_4)_2CO_3$(aq)

19 Cu^{2+}(aq) and Zn^{2+}(aq)

20 Pb^{2+}(aq) and As^{3+}(aq)

21 Al^{3+}(aq) and Zn^{2+}(aq)

22 Ca^{2+}(aq) and Mg^{2+}(aq)

Questions 23–26

Choose from the list A–E the reagent most suitable for carrying out each of the conversions specified in questions 23–26.

A I^-(aq)
B OH^-(aq)
C O_2^{2-}(aq)
D H_2(g)
E $LiAlH_4$(ether)

23 Cu(II) → Cu(I)

24 $Cr_2O_7^{2-} \rightarrow CrO_4^{2-}$

25 $Co(NH_3)_6^{2+} \rightarrow Co(NH_3)_6^{3+}$

26 $BCl_3 \rightarrow B_2H_6$

Assertion-reason questions

	ASSERTION		REASON
27	In titrating dilute hydrochloric acid with aqueous sodium hydroxide the acid should be put in the burette	*because*	sodium hydroxide may be pipetted with greater safety than dilute hydrochloric acid.
28	Potash alum may be crystallised either as octahedra or as cubes	*because*	both these shapes conform to the cubic system to which potash alum belongs.
29	Sucrose will burn more readily in air if it is finely divided than if it is composed of larger particles	*because*	when finely divided, a greater surface area is exposed to atmospheric oxygen.
30	Hydrochloric acid is used as a primary standard in volumetric analysis	*because*	hydrochloric acid is a monobasic acid.

17 Revision Paper I

Multiple choice questions

1 Which one of the following has the lowest bond energy?

A H—H
B Cl—Cl
C Br—Br
D I—I
E C—C

2 It is known that when a substance behaves as a Lewis base, the reaction normally takes place in such a direction as to remove the unpaired electron(s) from the least to the most electronegative atom. On the basis of this "rule", which one of the following reactions is likely to occur in the direction written?

A $CH_3COOC_3H_7 + Cl^- \rightarrow CH_3COO^- + C_3H_7Cl$
B $CH_3CN + Cl^- \rightarrow CH_3Cl + CN^-$
C $(CH_3)_3NH^+ \rightarrow (CH_3)_3N + H^+$
D $ICN + I^- \rightarrow CN^- + I_2$
E $CH_3O^- + CH_3Cl \rightarrow CH_3OCH_3 + Cl^-$

3 A reaction is first order with respect to a given reactant. If the initial concentration of this reactant is doubled, by what factor is the rate of the reaction affected?

A 1
B 2
C 4
D 8
E 16

4 When one Faraday of electrical charge is allowed to decompose dilute sulphuric acid, how many moles of oxygen gas at s.t.p. are liberated?

A 0.25
B 0.50
C 1.00
D 1.50
E 2.00

5 In which one of the following reactions is the metal oxidized?

A $Ni(g) + 4CO(g) \rightarrow Ni(CO)_4(g)$
B $Cr_2O_7^{2-}(aq) + 2OH^-(aq) \rightarrow 2CrO_4^{2-}(aq) + H_2O$
C $2Cu^{2+}(aq) + 4I^-(aq) \rightarrow 2CuI + I_2$
D $MnO_4^{2-}(aq) + HSO_3^{2-}(aq) + OH^-(aq) \rightarrow MnO_4^{3-}(aq) + SO_4^{2-}(aq)$
E $Bi(s) + 6HNO_3(aq) \rightarrow Bi(NO_3)_3(aq) + 3H_2O(l) + 3NO_2(g)$

6 Which one of the following will reduce aqueous mercury(II) chloride?

A HCl
B $SnCl_4$
C $PbCl_4$
D $SnCl_2$
E $PbCl_2$

7 Which one of the following reagents in aqueous solution effects the change

$$Al(s) \rightarrow AlO_2^-(aq) \quad ?$$

A NaOH
B $KMnO_4$
C SO_2
D H_2SO_4
E HNO_3

8 Which one of the following statements is correct?

A The boiling point of hydrogen sulphide is higher than that of hydrogen selenide.
B Germanium is expected from its position in the periodic table to form an oxo-salt of formula K_2GeO_4.
C The first ionization energy of sodium is less than that of rubidium.
D Iodine forms a series of oxides which are more stable to heat than those formed by either chlorine or bromine.
E Manganese is unknown in the oxidation state of 6+.

9 Which one of the following species is *not* of importance in the commercial extraction or purification of the metal concerned from its ores?

A $Ag(CN)_2^-$
B $Cu(NH_3)_4^{2+}$
C AlF_6^{3-}
D $Ni(CO)_4$
E Fe_2O_3

10 Which one of the following pairs of elements exhibit "a diagonal relationship" to each other?

A tellurium and iodine
B fluorine and sulphur
C beryllium and aluminium
D potassium and argon
E lithium and aluminium

11 Which one of the following helps to explain the fact that the melting point of magnesium oxide is over twice that of sodium fluoride?

A Sodium fluoride is a more covalent substance than magnesium oxide.
B The radius ratio Na^+/F^- is less than the radius ratio Mg^{2+}/O^{2-}
C Magnesium oxide contains doubly charged ions whereas the ions in sodium fluoride carry single charges.
D The fluoride ion is smaller than the oxide ion.
E The magnesium-oxygen bond is less strong than the sodium-fluorine bond.

12 Which one of the following correctly characterises the unit cell of magnesium oxide which has a face centred cubic structure?

A one Mg^{2+} and six O^{2-} ions
B one Mg^{2+} and eight O^{2-} ions
C six Mg^{2+} and six O^{2-} ions
D four Mg^{2+} and four O^{2-} ions
E one Mg^{2+} and one O^{2-} ion

13 Which one of the following statements is correct?

A At its melting point, a solid changes into a liquid with the evolution of heat.
B A given mass of substance has less potential energy in the vapour state than it has in the solid state.
C It is not possible to pass directly from the solid to the gaseous state.
D If a gas behaved ideally, it would be impossible to liquefy it.
E The vapour pressure of a liquid is always increased by the dissolution of a solid in it.

14 In the radioactive decay

$$X \xrightarrow{\beta^-} {}^{207}_{82}Pb$$

what is the atomic mass of X?

A 204
B 205
C 206
D 207
E 208

15 Which one of the following substances can be resolved into optical isomers *and* be oxidized to form a compound which reacts with HCN?

A $CH_3OCH(C_2H_5)$ with CH_3 attached to the CH carbon
B $CH_3CHOH.CH_3$
C $(CH_3)_2CH.CHO$
D $CH_3CHOH.C_2H_5$
E $(CH_3)_2CHCH_2OH$

16 Which one of the following compounds shows both basic and acidic properties?

A $HOCH_2CH_2CH_2.COONH_4$
B $H_2N(CH_2)_6NH_2$
C $CH_3CHOH.CH_2COOH$
D CH_3COOCH_3
E $CH_3CHNH_2.CH_2COOH$

17 Which one of the following does *not* undergo nucleophilic attack with acetone?

A HSO_3^-
B NH_2OH
C CN^-
D $Ag(NH_3)_2^+$
E CH_3MgI

Assertion-reason questions

	ASSERTION		REASON
18	If an ionic substance dissolves in water endothermically, its lattice energy is greater than the heats of hydration of its constituent ions	*because*	heat energy is required to break down the ionic lattice and part of this energy may be supplied by the hydration of the ions.
19	$Al(H_2O)_6^{3+}$ is a stronger acid in aqueous solution than $Na(H_2O)_6^+$	*because*	the aluminium cation carries a triple positive charge whereas the sodium ion is unipositive charged.
20	Sulphur dioxide molecules have no net dipole moment	*because*	sulphur and oxygen are in the same group of the periodic table.
21	Moist starch potassium iodide paper is turned blue by the action of chlorine gas	*because*	chlorine oxidizes the iodide ions to iodine which reacts with the starch.
22	Increase of pressure has no effect on the equilibrium $N_2(g) + O_2(g) \rightleftharpoons 2NO(g)$	*because*	there are equal numbers of molecules of reactants and products.
23	The presence of a solid radium compound may be detected by means of a zinc sulphide screen	*because*	the decay of radium atoms produces ions in the surrounding atmosphere.
24	The halogens in the gaseous state are coloured	*because*	halogen molecules absorb some parts of the visible spectrum of light.

25 The vanadate ion VO_3^- is a powerful reducing agent in aqueous solution *because* vanadium is in the oxidation state expected from its group in the periodic table.

26 The bonds in carbon tetrafluoride are arranged tetrahedrally around the carbon atom *because* the bonding orbitals involve sp^3 hybridization.

27 Under given conditions hydrogen selenide has a lower boiling point than hydrogen telluride *because* hydrogen telluride is a heavier molecule than hydrogen selenide.

28 In the absence of oxygen, graphite remains a solid at 2000°C and 1 atmosphere pressure *because* planes of carbon atoms are held together by delocalised electrons.

29 Potassium permanganate is used as a primary standard in volumetric analysis *because* it is stable in air and in aqueous solution.

30 Methylamine is a liquid at room temperature and pressure *because* intermolecular hydrogen bonding is present in liquid methylamine.

18 Revision Paper II

Matching pairs questions

Questions 1–4

Choose, from the list A–E, the compounds which under appropriate conditions have the properties described in each of questions 1–4.

A $NaNO_2$
B $NaOH$
C NH_2OH
D N_2H_4
E NH_3

1 forms an oxime with benzaldehyde (benzenecarbaldehyde)

2 forms an alcohol and a carboxylate anion with benzaldehyde (benzenecarbaldehyde)

3 combines readily with acetic (ethanoic) acid to form a salt which on heating yields an amide

4 yields a phenol when added to *o*-toluidine (2-methylphenylamine) and dilute hydrochloric acid and heated

Questions 5–8

Choose from the list A–E the compound which correctly accounts for the properties stated in each of questions 5–8.

A CO_2
B $HCFClBr$
C $N(CH_3)_3$
D $CHCl_3$
E $CH_3CH_2CH_2Br$

5 has positional isomers

6 exhibits stereoisomerism

7 forms a salt with aqueous hydrogen chloride

8 has zero dipole moment

Questions 9–12

Choose from the list A–E the most appropriate description of the role of the water molecules in the reactions in each of questions 9 to 12.

A oxidant
B reductant
C proton acceptor
D proton donor
E electron pair donor

9 $HF + H_2O \rightarrow H_3O^+ + F^-$

10 $BF_3 + H_2O \rightarrow H_2O.BF_3$

11 $2F_2 + 2H_2O \rightarrow 4HF + O_2$

12 $H_2O + N_2H_4 \rightarrow N_2H_5^+ + OH^-$

Multiple completion questions

13 Which of the following conversions involve(s) oxidation?

(*i*) $MnO_4^{2-} \rightarrow MnO_4^-$
(*ii*) $MnO_2 \rightarrow MnO_4^-$
(*iii*) $Mn(H_2O)_6^{2+} \rightarrow Mn_2O_7$
(*iv*) $Mn_2O_7 \rightarrow MnO_4^-$

14 Which of the following substances dissolve(s) in 2M nitric acid?

(*i*) AgCl
(*ii*) AgI
(*iii*) Au
(*iv*) Ag_2CO_3

15 Which of the following will lead to a reaction involving disproportionation?

(*i*) addition of aqueous iodide ions to aqueous Cu^{2+} ions
(*ii*) addition of water to solid PCl_5
(*iii*) addition of acidified $MnO_4^-(aq)$ to aqueous H_2O_2
(*iv*) heating $OCl^-(aq)$

16 In which of the following reactions does the $HSO_3^-(aq)$ ion act as an acid?

(*i*) $HSO_3^-(aq) + H_3O^+(aq) \rightarrow H_2SO_3(aq) + H_2O(l)$
(*ii*) $HSO_3^-(aq) + CH_3CH_2NH_2(aq) \rightarrow CH_3CH_2NH_3^+(aq) + SO_3^{2-}(aq)$
(*iii*) $HSO_3^-(aq) + HCl(aq) \rightarrow H_2SO_3(aq) + Cl^-(aq)$
(*iv*) $HSO_3^-(aq) + H_2O(l) \rightarrow H_3O^+(aq) + SO_3^{2-}(aq)$

17 In the light of the following equilibrium, for which $K > 1$,

$$CH_3CH_2COO^-(aq) + H_2O(l) \rightleftharpoons CH_3CH_2COOH(aq) + OH^-(aq)$$

which of the following statements is/are correct?

(*i*) CH_3CH_2COONa will be acidic in aqueous solution
(*ii*) $CH_3CH_2COO^-(aq)$ is a stronger base than $OH^-(aq)$
(*iii*) The water is acting as a Lewis base.
(*iv*) The addition of aqueous alkali will move the equilibrium to the left.

18 The dissolution of ammonium chloride in water is accompanied by the absorption of heat. Which of the following statements is supported by this experimental fact?

(*i*) The process of solution reaches equilibrium only when the solution is saturated.
(*ii*) Reactions proceed to a position of minimum energy.
(*iii*) The solubility of ammonium chloride in water decreases as the temperature is raised.
(*iv*) The dissolution of ammonium chloride in water is an example of a spontaneous endothermic reaction.

19 Which of the following sets of reactants may produce gaseous hydrogen?

(*i*) $NaH(s) + H_2O(l)$
(*ii*) $NaNH_2(s) + H_2O(l)$
(*iii*) $Cu(s) + H_3O^+(aq) + I^-(aq)$
(*iv*) $Cu(s) + H_3O^+(aq) + NO_3^-(aq)$

20 Which of the following statements about the emission spectrum of hydrogen is/are correct?

(*i*) The spectrum of a given series consists of a set of lines which eventually converge.
(*ii*) A value for the ionization energy of hydrogen can be obtained from the spectrum.
(*iii*) The lines arise from electron transitions between one energy level and another.
(*iv*) The number and/or position of lines in the spectrum cannot be affected by external factors.

21 Which of the following statements about spectra is/are *incorrect*?

(*i*) An argon filled electric light bulb will give a continuous spectrum when viewed through a spectroscope.
(*ii*) The spectrum of the sun is a continuous spectrum containing numerous dark lines.
(*iii*) Any sodium compound in a bunsen flame will give a spectrum characterised by two yellow lines occurring close together
(*iv*) A fluorescent lamp will give a continuous spectrum when viewed through a spectroscope.

22 Which of the following properties would you predict for sodium bismuthate $NaBiO_3$?

(*i*) It will liberate chlorine from concentrated hydrochloric acid.
(*ii*) It will liberate chlorine from dilute hydrochloric acid.
(*iii*) It will hydrolyse in water.
(*iv*) The passage of dry hydrogen gas over the heated solid yields NaH, BiH_3 and H_2O.

23 Which of the following species contain(s) catenated atoms (*i.e.* atoms of the same element linked together)?

(*i*) H_2O_2
(*ii*) $H_4P_2O_7$
(*iii*) C_2H_4
(*iv*) H_3PO_4

24 Which of the following properties shows an increase in every group (or sub-group) of the periodic table as the group (or sub-group) is descended?

(*i*) atomic number
(*ii*) electronegativity
(*iii*) first ionization energy
(*iv*) atomic mass

25 Which of the following elements do/does *not* form a well characterised hydride?

(*i*) arsenic
(*ii*) mercury
(*iii*) potassium
(*iv*) xenon

26 Which of the following properties would you expect of the element radium?

(*i*) Its principal oxidation state is +2.
(*ii*) Its hydroxide is appreciably soluble in water.
(*iii*) Its sulphate is insoluble in water.
(*iv*) It forms a stable, solid bicarbonate.

27 Which of the following have sulphates which are effectively insoluble in water at room temperature?

(*i*) lead
(*ii*) lithium
(*iii*) calcium
(*iv*) aluminium

28 Which of the following substances yield at least one gaseous *compound* of hydrogen when treated with dilute hydrochloric acid?

(*i*) Mg_3B_2
(*ii*) Mg_2Si
(*iii*) CaC_2
(*iv*) KH

29 Which of the following elements would you reasonably expect to be extracted from its ore by a process involving electrolytic reduction?

(*i*) iron
(*ii*) potassium
(*iii*) bismuth
(*iv*) barium

30 Which of the following statements about glucose and sucrose is/are *false*?

(*i*) Both are naturally occurring.
(*ii*) Both have isomeric forms.
(*iii*) Both are crystalline solids under normal conditions.
(*iv*) Both are disaccharides.